DH88

de Havilland's RACING COMETS

DH88

de Havilland's
RACING COMETS

David Ogilvy

Airlife
England

ISBN 1 85310 011 0

This edition published 1988
by Airlife Publishing Ltd.

First hardback edition published 1984
by Airlife Publishing Ltd

Printed in England by Livesey Ltd., Shrewsbury.

Airlife Publishing Ltd.

7 St. John's Hill, Shrewsbury, England.

Contents

Acknowledgements

It is impossible to list here every person who has made a contribution towards the facts that I have endeavoured to assemble. In the Introduction I have referred to those who sent information from far afield, but nearer home many others have devoted considerable time, energy and patience to provide the necessary detail. These include Stan Baker, a de Havilland engineer in 1934; Bruce Bosher, Product Support Planning Manager at Rolls-Royce, Leavesden; Hugh Buckingham, who completed many hours of Comet flying as the de Havilland sales and demonstration pilot; R. W. (Bob) Head, Senior Engineer, Research, at British Aerospace, Hatfield; Richard Clarkson, who flew as test observer with Hubert Broad on many test and development flights; Mrs. C. F. Hughesdon (formerly the actress Florence Desmond, who married Tom Campbell Black shortly after the MacRobertson race); Peter Murton, Librarian and Ken Hunter, archivist at the Royal Air Force Museum; and John Thorpe of the Civil Aviation Authority.

Among those associated directly with the Shuttleworth Collection who have provided much of the meat for the recent restoration story I mention Ron Paine, the honorary Project Co-ordinator; and Chris Morris, the Collection's Chief Inspector. Each, too, has provided pictures, as has Jocelyn Millard, the Collection's Information Officer.

Photographs have come also from others, all of whom I wish to thank. These include *Flight International* (Business Press International), British Aerospace, David Davies and Michael Vines of *Air Portraits,* John Hoad and especially Richard Riding, editor of *The Aeroplane,* whom I have known for nearly thirty years as an avid purveyor of pictures.

At one stage I became confused by the amount of conflicting information that I was unearthing and I lost myself plunging into several blind spots. I am not trained in research, but Audrey, my wife, is very experienced in this field and I wish to publicise my appreciation of the way in which she redirected my thoughts back onto a straight and level heading.

These are some of the people who have played their parts specifically in the preparation of this book. Those who have helped in other ways, especially in the restoration project, are mentioned in the text or in appendices. If I have omitted any, I apologise, but whether mentioned or not, thanks are due to them all.

Introduction

At the end of 1966 I became involved with the Shuttleworth Collection at Old Warden Aerodrome in Bedfordshire. Not long passed before I began to feel uneasy each time I walked past one of the most famous aeroplanes on display, knowing that it was one of the very few not in flying condition. Richard Shuttleworth, the Collection's founder, had decreed from the start that everything should work at least as well as it had done when it was built.

This famous aeroplane was the de Havilland DH88 Comet G-ACSS, which had been restored externally to its original colours as *Grosvenor House*, as which in 1934 it had won the MacRobertson Air Race from Mildenhall in England to Melbourne in Australia. It was the sole complete survivor of the five Comets built and it had a very special place in the history of British aviation. Several people agreed that it should fly again, although for a while a few tried determinedly, but with the best intentions, to keep it permanently grounded. Eventually the proposal to make G-ACSS airworthy again was agreed by all.

I have endeavoured to tell the tale from 1934 to today. However, I could not have completed this without considerable help from others and I am grateful to those who responded to requests for information that were published in at least three aviation magazines. I am sure that their contributions have helped considerably towards making this the truth, the whole truth and, I hope, nothing but the truth.

Some readers may challenge some of the statements of fact, but here I must explain a problem. Before I started my research into the Comet story, I thought that I knew most of it. I was quite wrong. The Comet first flew only fifty years ago, yet much information has been lost and many records have become confused. As an example, I was very surprised to find difficulty in establishing which of the original three Comets was the first to fly. One well-known historian recorded this as G-ACSP while another, of equal note, stated it to be G-ACSR. This set me searching further. As a result, I delved over a period of nearly three months, during which time one established source confirmed categorically that it was G-ACSR and then backed this by adding the constructor's number, which was that for G-ACSP! The problem has been compounded throughout because for the first few flights no registration letters were carried and the only markings displayed, E-1, have been used on at least four aeroplanes. In the end, I settled for an entry in the personal logbook of

Hubert Broad, then de Havilland's chief test pilot, who is known indisputably to have carried out the first flight. Copies of the relevant pages are printed on page 26.

Some readers may feel that this is only a minor point, but I quote it as one of several problems that have increased in difficulty as the book has advanced. I have aimed for accuracy of fact and detail, but I am ready for brickbats from those who find errors or omissions. At the start, I *did* ask for help, which as a result has come from more than twenty people as far afield as Australia, South Africa, the United States and Canada. However, if anyone can offer any facts that I have missed, in the optimistic hope that this may run into a second edition, I would like to hear. Letters should be sent to the publisher, please.

Finally, I must make clear that this book is a purely private venture and although I could not have written it without a close insight into the workings of Old Warden, I am not expressing official Shuttleworth Collection policy. The pages that follow contain my own findings about the whole Comet story, from the first thoughts of the race to Australia until mid-1984.

David Ogilvy
Summer 1984

Chapter 1
Why it all happened

Except during the last two years of the decade, the aeroplanes of the nineteen-thirties were similar in general layout to those that had flown in the 1914-18 war; the high-speed fighters of the time were Hawker Furies and the bombers ranged in size and weight from Hawker Harts to Boulton Paul Overstrands and Vickers Virginias, while the sporting mounts were such types as the Avro Avian and various versions of the ubiquitous de Havilland Moth. Most were biplanes, with fixed undercarriages, open cockpits and fixed-pitch propellers.

Two people, more than eleven thousand miles apart, were to change this. Australia was in the midst of a serious slump and in need of closer and more rapid contact with Great Britain; nowhere was this need felt more forcefully than in Melbourne, Victoria, once the country's capital and leading city, but which had lost these symbols of status to Canberra and Sydney respectively. Many people had moved away from Melbourne, which had suffered a succession of turbulent times; these had ranged from the riches of the gold rush in the 1850s to a general move to farming and manufacturing, with a resulting series of major bankruptcies. All this trouble was followed by about twenty years of relatively stable prosperity, but after several more decades of instability the slump had hit hard.

The one-hundredth anniversary of the foundation of the State of Victoria would fall in 1934, so two years before this a Festival Committee was formed to plan the inevitable civic celebrations. All the usual events were to be staged, but allegedly the Lord Mayor, Sir Harold Gengoult-Smith, considered these to be far too parochial and he is credited with proposing the idea of an air race, although more recent research places doubt about whether he conceived the plan or just supported it. However, the truth about this is relatively unimportant, for he approached Sir Macpherson Robertson, a millionaire manufacturer of sweets, to ask if he would sponsor such an event. Subject to some rigid requirements regarding the safety side, the rich man reacted quickly and favourably to the idea; he saw this largely as a means of reducing the geographical isolation from which Australia was suffering.

Sir Macpherson had seen both sides of the coin of fortune. He was born in Australia in 1860, but his parents were Scottish; when he was very young the family returned to their homeland to live in poverty there before going back to Australia when the lad was fourteen years of age. Young Macpherson was determined to become a confectioner and in 1880 he established a small business in the family bathroom. By the age of 25 he had his own premises with a staff of fifteen; five years later his empire was waxing steadily with a growing chain of factories. He was intensely patriotic about Australia and, among other ventures, he had financed the 1929-31 joint Australian and New Zealand Antarctic expedition led by Sir Douglas Mawson. In 1932, at the age of 71, he was knighted. Throughout this time he had preferred the use of a contraction of his full name and he became known to most people as 'Mac Robertson'.

On 1st April 1932 plans were agreed for a race committee to be formed under the chairmanship of the Lord Mayor of Melbourne. However, a year and two days passed before the first formal meeting, at which Sir Macpherson Robertson delivered this charge:

> 'You have been convened to make arrangements for an International Air Race from London to Melbourne in connection with the Centenary Celebrations. Make it the greatest race yet conducted in the world; make as few conditions as possible consistent with reducing risks to the minimum. Give attention to Australia's geographical isolation, and in this connection direct your thoughts towards the possibility of reducing this handicap consistent with safety.'

This was a major challenge to all present and, with the event only eighteen months ahead, little time was available for detailed consultation. The committee was not brimming with practical aviation expertise, but for a guideline members reviewed the events of the first race from England to Australia that had taken place in 1919. Then five aircraft and crews had competed for a prize of £A10,000 offered by the Prime Minister to any Australian citizen(s) who would complete the journey in thirty days, providing the result was achieved before the last day of that year. The outcome had not been wholly encouraging, as four of the five participating aeroplanes had come to grief on the way, leaving Captain Ross Smith and his brother Lieutenant Keith Smith, with two sergeant fitters, as the only crew to complete the course, in a Vickers Vimy powered by a pair of Rolls-Royce Eagle engines.

The plans for the 1919 event had been necessarily sketchy; basic facilities (including petrol by Shell and oil by Wakefield) were laid on as far as Calcutta, which had no aerodrome so the race course had been brought into use, but beyond there not even fuel supplies would be guaranteed. However, success was made possible by the co-operation of the Air Ministry, with RAF personnel laying on whatever help they were able to offer. Captain Ross Smith and crew reached Darwin 27 days and 20 hours after departure from England, later (unknowingly) pre-empting the 1934 race by flying on to Charleville and Melbourne, but the problems experienced en route gave much scope for thought and worry by the newly-formed committee: the Vimy had scraped trees after a fully-laden take-off at Rangoon; at Surabaya (the next stop after Singapore) the aircraft had become bogged down following tropical rain storms and had been able to leave again only after hundreds of coolies had laid matting; and the climb-out after taking off from Timor Island necessitated weaving turns at low level to avoid trees.

Interestingly, there were reports of Captain Smith convincing himself that his Vimy was suffering from engine-running problems when over long stretches of water (and the Timor Sea offers about 900 miles of it) but in practice these were psychological rather than real problems. This is a form of fear that inflicts itself on many pilots even today. After it all, deservedly, both the Smith brothers were knighted and the two NCOs were awarded Air Force Medals.

There were many subsequent achievements — and several failures — for the committee to study as a part of the preparation process. Several individual pilots had attempted (and some had succeeded in) covering the distance of more than 11,000 miles, usually trimming time from their predecessors' achievements to establish new records. In 1928 the Australian Bert Hinkler had flown an Avro Avian single-engine light biplane from Croydon to Darwin in little more than half the elapsed time taken by the Vimy and, within a year, Charles Kingsford Smith had achieved the distance in 13 days, reducing it in a later attempt to only ten days. In 1930 Amy Johnson had become the first woman to achieve the journey solo.

There were many subsequent flights to and from Australia and one that must warrant mention is the successful attempt by Alan Butler in a tiny single seat Comper Swift. This little Pobjoy-powered high-wing monoplane had a cramped cockpit, virtually no luggage capacity, a wing in the way of forward view and an engine that tended to eject blobs of oil on windscreen and goggles on any flight of more than short duration. In the fifties I had the good fortune to fly one such machine in several relatively short air races within the comfortable confines of the British coastline and I cannot imagine many machines that would offer less ease and convenience on long and tiring legs over unknown and often inhospitable territory. We should think hard about the achievements of these pioneers.

As far as this book is concerned, perhaps the most significant of all the successful flights between the two countries may be those by C .W. A. Scott in a Gipsy Moth from Lympne to Darwin in April 1931 (when he completed the route in 9 days 3 hours and 40 minutes) and another a year later in which he trimmed the time to just under 9 days; but we must not overlook the achievement by James (Jim) Mollison who in July 1931, also Moth-borne, reversed the process by flying from Sydney on 25th July and arriving at Pevensey on the south coast of England in 8 days 22 hours and 15 minutes. The importance of these two names among the earlier achievements will come to the fore later.

Whilst many of these records may not have very direct bearings on the main subject of this book, they were of considerable importance in two specific spheres: firstly, by studying the stage lengths, the support problems, the difficulties caused by hazardous terrain and the stresses imposed upon both the aircraft and their pilots, the fledgling race committee in Melbourne could gain much of the knowledge needed for the early organising stages of the great race to come; and secondly, prospective participants could learn some of the facts to be faced when committed to flights over such extreme distances.

Although much more detailed planning was to follow, Sir Harold Gengoult-Smith and his team realised that time was running out, especially if new aircraft were to be designed and built or developed specifically for the race, so they published the basic requirements for competitors: there would be no limit to the size or power of participating aircraft and no restriction on the permitted number of operating crew, but to qualify, all aircraft must fly the entire route from England to the laid-down finishing point at Flemington race-course outside Melbourne. Also at this stage some preliminary rules were drafted regarding minimum food, protective equipment and smoke/safety signals to be carried.

Whilst all this was engaging the attention of the committee members, Mac Robertson was seeking all possible ways of increasing airmindedness in Australia. The first flight had taken place in March 1910 and in the 1914-18 war there had been limited activity by the Australian Flying Corps, while the first commercial air service had been provided by Larkin Aircraft Supply Company in 1921 between Adelaide and Sydney. Six years later Mac had financed the Commercial Aviation Company; by 1932 there were 53 privately licensed and 58 Government aerodromes spread throughout Australia, but there was far too little flying and there were far too few pilots for aviation to be considered a healthy and firmly established activity. Although civil operations were the main target in the benefactor's mind, he was not overlooking the growing need for military air strength caused by troubles with the Japanese to the north.

Suprisingly, possibly through ignorance of the international aviation sporting code, the Melbourne race committee kept their more formal thoughts and their intentions very largely to themselves, for not until 20th April 1933 did the secretary write to the Royal Aero Club in London to seek that body's help or blessing for the event. I should add here that the club (formed in 1901 initially to foster the sport of ballooning) has been actively involved throughout in organising or approving all forms of air racing in Britain and, through its affiliated Records, Racing and Rally Association continues to do so today. It acts as the United Kingdom representative on the Fédération Aéronautique Internationale, which is the worldwide organisation for monitoring and accepting all forms of records, so to think how the Australians hoped for success without Aero Club involvement leaves much scope for doubt about the depth of the early planning.

Available information indicates that the Royal Aero Club met on 26th April 1933 to consider the race, which would have been impossible if the letter had followed the conventional postal pattern, so perhaps its contents had been sent to London by telegraph. However, in whatever way the news was received, the club moved with unaccustomed speed and almost immediately some positive proposals emerged. Fortunately Air Commodore R. Williams, the Australian Chief of the Air Staff (whose rank alone supported Mac Robertson's concern about the small scale of military air activity there) was in London and was appointed as liaison officer between the two countries.

The route was established, with compulsory stops for refuelling and checking (and for crew convenience) at Baghdad, Calcutta, (changed on 1st March 1934 to Allahabad), Singapore, Darwin and Charleville before flying to the finish at Melbourne; these were known as control points, but there were several additional and optional checking/refuelling points along the way. To offer opportunities for a wide range of aircraft types, a proposal from Australia that the race should be sub-divided into open and handicap sections was accepted in London; but time spent at the optional places would count as flying time, whilst time at the compulsory control points spent eating, sleeping, servicing and refuelling would not be recorded as such.

Detailed requirements included a closing date for entries of 1st June 1934, with accompanying fees of £50 for the open section and £10 for the handicap, returnable in full for each aircraft that took off to begin the race. Each participating aeroplane was to be at the starting point seven days before the event, which was planned for Saturday 20th October or as near as possible to that date. Aircraft and engines would be sealed one day before the start and no major replacements would be allowed, with the seals to remain unbroken for final checking after the race. Surprisingly, both night flying and flight refuelling were fed into the rules as permitted activities.

Not surprisingly, though, disagreements reached fever levels. The Australians wanted the open and handicap sections to run simultaneously and the Royal Aero Club pressed that they should be flown as two separate events. These and other equally significant matters were subjects for hot argument until almost the end of the year, with two members of the organising committee resigning as late as November. The friction was not restricted to relations between England and Australia, for the United States entered the fray with considerable opposition to some of the British proposals, on the basis that they were geared to the needs and wishes of the most likely contenders from the United Kingdom.

Despite all the international and internal bickering, the plans developed and Mac Robertson continued to look ahead to the race as an established entity. After earlier, smaller offers of prize money, he settled for the final figures of £A10,000 for first prize in the open race (with second and third awards of £A1500 and £A500 respectively) and £A2000 (£A1000 for second place) in the handicap section. Also, the winner would receive a gold cup valued at £A650. A firm ruling from the beginning, though, was that regardless of physical results no one competing aircraft and crew could qualify for both the open and handicap prizes. Although attractive, the prize money was not so great as it may sound, as at that time £A10,000 would have been worth about £7,500 in sterling.

The main route details were finalised at the first special meeting of the Royal Aero Club, but political problems persisted until alarmingly late dates. For many months after being requested, Turkey failed to grant permission for competitors to overfly that country; when eventually the authority was given, the situation was partially worsened, as it was offered only to British participants, who could fly over or land there as they wished, but pilots and crews of all other nationalities would need to seek and obtain special individual clearances. There were difficulties nearer home, too, for initially no suitable starting aerodrome could be found!

Although earlier in this chapter I have referred to several of the pioneers who had worked their individual ways in the air between England and Australia, all at a time when international commercial air travel was almost unknown, I must not omit the unique efforts of one organisation that stood ahead of the others: the Dutch airline KLM, formed as a private company in October 1919 by Albert Plesman, who steered his firm in positive direct drive throughout. In the early twenties he had planned a regular scheduled service from Europe to the Far East, but he had been unable to fulfil this aim through lack of suitable aircraft for the job. However, he persuaded the Dutch Government to establish a Netherlands East Indies Flight Committee, which eventually sponsored a trial flight in a Dutch-designed and Dutch-built Fokker FVII to Djakarta in the (then) Dutch East Indies. Although this first attempt at long-range travel was problem-plagued, later flights to and from the same destination proved trouble-free. An agreement was reached with the Netherlands Post Office for air mail contracts and Plesman established airfields and refuelling facilities at strategic points along the routes to be developed. At this stage perhaps KLM was the most ambitious airline in existence anywhere and I mention it now because of its involvement in the events to follow.

As the day of the race grew closer, more details were published, including a formula for computing the handicaps for the various expected entrants. This was based on payload in lbs (allowing 200 lbs for each occupant), total weight in lbs, the maximum horse-power of the prototype engine(s), (with different requirements for those rated at sea level and those that were 'blown' for rating at higher altitude) and the area of the mainplanes (wings). Again expectedly, even the formula became the subject of internal disputes.

Considerable research and effort were needed to find and guarantee the ready availability of some of the optional checking points, which would be essential to enable participants flying relatively short-range aircraft to have reasonable chances to complete the course in acceptable safety. There were no problems over the first three, at Marseilles, Rome and Athens, for they were established aerodromes. The remainder, at Aleppo, Bushire, Jask, Karachi, Jodhpur, Calcutta, Rangoon, Bangkok, Alor Starr, Batavia, Rambang, Koepang, Newcastle Waters, Cloncurry and Narromine, called for help from various bodies, including the Royal Air Force, the Civil Aviation Department in India, KLM and the Royal Australian Air Force.

Because of the complexities of the race and the vast sums of money that were to be put on risk, an insurance pool was formed, comprising all the leading underwriters and companies that handled aviation cover. For any aeroplane valued at under £1,000 comprehensive protection was offered for a premium of 20 per cent of the insured value; the rate was 15 per cent for aeroplanes assessed between £1,000 and £10,000 and the Comets were included in this range, while for the machines above this figure the charge was 10 per cent.

However, in the event of no call being made on the policy, a no-claim benefit of one third of the premium was returnable. This scheme included the aircraft hull and, of course, third party liabilities, but for any pilot who wished to cover his or her own life this was offered at 8 per cent for death only or 10 per cent to cover death or injury. There was no refund on this life policy.

By this time Mildenhall had been proposed as the starting point. The distances to be covered between the compulsory stops would be far too great for many of the competing aeroplanes, but for participants wishing to make maximum use of total elapsed times, the fewer the landings and refuellings the better their chances would be, so some machines were expected to be developed to accommodate overload fuel tanks. At the same time, though, the regulations required all aircraft to be able to operate within the terms of the existing categories of their Certificates of Airworthiness. Those entrants planning to fly the route with the minimum of delays, therefore, would be required to cover these long legs and still have sufficient fuel reserves to cope with navigational errors or setbacks through bad weather or both. These distances were:

Mildenhall to Baghdad	2,530 miles
Baghdad to Allahabad	2,300 miles
Allahabad to Singapore	2,210 miles
Singapore to Darwin	2,084 miles
Darwin to Charleville	1,389 miles
Charleville to Melbourne	787 miles

The shorter legs at the end were intentional, for by then crews and aircraft would be suffering from severe fatigue.

Rules, regulations and requirements for the race were becoming firm and entries were open, but what aircraft would be taking part? Participants were expected from many parts of the world, including Australia, New Zealand, the Netherlands and the United States, but which established types of British design and construction would be likely contenders in an event that was showing signs of becoming the most important contest ever to be held in the history of aviation? Suitable machines with real prospects, alas, proved to be missing from the Register, but surely Great Britain could not lie flat in this moment of global challenge.

The details must await the next chapter, but from the efforts of Sir Macpherson Robertson in Australia and the enterprise of Geoffrey de Havilland in England, a very special aeroplane emerged: the de Havilland DH 88 — the first Comet, which is the whole and sole reason for this book to be written — so read on . . .

Chapter 2
The Comet Appears

By the beginning of 1934 no offer of a new design of aircraft specially for the MacRobertson Air Race had emerged from the British aircraft industry. A. V. Roe had declined to show any genuine interest, while Handley Page, Vickers and Boulton and Paul went through various discussions and indulged in some preliminary paper exercises. However, one by one they had dropped out of the running, leaving only one serious home contender in the form of the de Havilland Aircraft Company; but this firm was very serious. From the outset Geoffrey de Havilland had shown a determination to design and build a winner and the eventual result grew from his initial comment "You know, we can't stand by and let this race be run without any British effort".

In January 1934 ,when design work began, the original idea was to develop a racer in the form of a larger twin-engine development of the DH71, a small wire braced monoplane with a fixed undercarriage that in 1927 had achieved 186 mph on only 135 hp. — the first of two DH types to carry the name Tiger Moth. However, ideas changed quickly and the team, under A. E. Hagg as chief designer, set to work on a low-wing twin-engined cantilever monoplane with enclosed cockpit, flaps and a retractable undercarriage. With only nine months to go before the start of the race, this task alone would have daunted most people, but already there were further problems to be solved.

Fortunately, de Havilland's own Gipsy Six, six-cylinder, air-cooled, in-line engine was developing successfully as a big brother to the already-established four-cylinder Gipsy Major and, in principle, this appeared to be eminently suitable for the new aeroplane. It had the potential for an increase in power output from its published figure of 185 bhp, but to use this effectively both for take-off at heavy weight and for economical long-range cruise at high speed, the blade setting for any fixed-pitch propeller would be too much of a compromise to provide a worthwhile performance in either case. No suitable variable-pitch propeller existed in Britain and the de Havilland team needed to look far to find one that might fill the bill; but more of that later.

Another major headache was the method of construction, especially for the mainplane. Geoffrey de Havilland had anounced very little about the new design, except that it would be capable of exceeding 200mph and could tackle stage-lengths of at least 2600 and possibly 3000 miles, so with only that bare information to offer to prospective purchasers, the company placed this brief advertisement in *Flight* and *The Aeroplane*; the news was circulated also in Australia:

> 'The de Havilland Comet, now being designed for the MacRobertson England to Australia International Air Race. Orders are invited for a limited number of this long-distance type of racing aircraft.'

The advance but essential commitment to both high speed and long-range demanded an aeroplane of very low frontal area and very high aerodynamic efficiency. This meant not only that the wing would need to be too thin to accommodate any fuel tanks, but its depth would be insufficient for conventional load-bearing spars. This necessitated an excursion into a new method of stressed-skin construction; in this the spars would carry the shear loads and transmit loads to the skin, which would carry the bending and torsional stresses.

Construction would be entirely of wood, with shallow box spars joined by a skin comprising two layers of spruce planking, crossing each other at 60 degrees diagonally to the fore-and-aft axis of the aircraft. This crossed planking had been used in boat-building and a simplified pattern had been incorporated in the French Deperdussin racer as early as 1910, but in its present form it was a new venture in the aviation world. The strips were two inches wide throughout, but the combined planks tapered in thickness from ·5″

The first DH 88 Comet on assembly. The aircraft suspended from the roof is one of the two DH 71 racers built in 1927. (The Aeroplane).

at the wing root to ·14″ near the tip. To allow for cut-outs through this skin in the centre section, two additional layers of ·07″ planking were added outward to about four feet either side of the fuselage centreline. From about six feet inboard of the tips, to the tips themselves, various thicknesses of ply were used because of the difficulty of machining spruce more thinly than ·07″. The full length of the leading edge (ahead of the front spar) was ply covered, with a spanwise stringer for support. Diagonal spruce planking was used also for the upper and lower surfaces of the fuselage forward of the cockpit in order to obtain the very critical double curvature required to provide the clean lines, for although ply sheet can be bent to form a cylinder or a cone, it cannot be formed into the shape of a barrel. As the fuselage sides were flat, these were covered in normal plywood, as were the top and bottom surfaces of the rear part of the fuselage.

According to the original drawings the three continuous box wing spars were made of 'American Whitewood' webs with spruce flanges, but in practice the webs appear to have been made from birch ply (and certainly this has been used in the restoration programme described in Chapter 9). These spars were at 21, 40 and 65 per cent of the chord with a spruce stringer added between each pair of spars to give support to the planking to prevent buckling. The ribs were built from birch ply with spruce uprights, corner blocks and capstrips. The long, thin ailerons, which had mass balance weights inside the leading edges, were ply covered.

Quite apart from its unique structural features, the Comet wing was unusual aerodynamically. The very thin low-drag section with a slightly reflex trailing edge (RAF 34) had a very high lift/drag ratio and to obtain the maximum benefit from this feature the incidence setting was less than one degree. Not only was the wing thin in depth, but also in chord, with an aspect (i.e. span to chord) ratio of 10:1. The reflex trailing edge was continued into the large curving sheet metal root fillet that smoothed the airflow between the wing and the fuselage.

The integral fin and the one-piece tailplane were constructed with two light solid spars and plywood ribs that were similar to those used in the wing; they also were ply covered. The rudder and elevators were built similarly and each had bob-weight mass balances. The flying control runs included standard de Havilland differential units to prevent over-sensitivity at high speed and excessive control column movements at the low end of the scale.

With the entire design concept geared to speed and range, the fuselage needed to accommodate a large supply of fuel, with three tanks holding a total of 258 gallons. The smallest tank was of only 20 gallons capacity, placed just behind the cockpit where with a long moment from the centre of gravity, it could help with longitudinal trimming on long legs as the main fuel supply was used. To keep frontal area to a practical minimum the narrow cockpit had a flatly sloping windscreen and was positioned well aft, providing only cramped accommodation for the two tandem-seated occupants and offering a poor view directly ahead. Flying controls were duplicated, but flight instruments were provided for the front position only.

To obtain the promised performance, the team needed to venture for the first time into the design and construction of a retractable (or as it was called, retractile) undercarriage. This was a very basic device controlled by a handwheel in the cockpit and connected to endless cables; these, via a helically grooved drum with several wraps of the cable and worm form actuated a screwed shaft to raise and lower the wheels. A mudguard was fitted over the front of each wheel and when the undercarriage was retracted this formed part of the bottom of the engine nacelle.

I have mentioned the fortunate availability of the up-and-coming Gipsy Six as the source of power, but even this required considerable development from its basic version. Cylinder heads were changed from bronze to aluminium alloy units and, to raise the compression ratio from the standard figure of 5.25:1, metal was removed from the cylinder head flange; however, on the first tests detonation resulted, but as cast pistons with domed (instead of flat) heads had been developed for racing versions of the Gipsy Major these were used in the special engine, which became known as the Gipsy Six R. Other changes included modification to the valve gear. The result was an engine with a compression ratio of 6.5:1 and a considerable increase in performance. Many discussions were held regarding the pros and cons of using supercharged engines. These would have cleared the Comet to take-off at a higher loaded weight therefore enabling more fuel to be carried, but Frank Halford, the designer of the Gipsy Six, decided that the increased rate of fuel consumption would have cancelled this benefit, so normally aspirated units won the day.

Power output from an engine is difficult to determine on a comparative basis, for figures vary with the different settings and measurements used. Although the basic Gipsy Six 1 was credited with an output of 185bhp at 2100rpm for take-off, elsewhere an output as high as 205hp has been stated. However, quoting from the 185/2100 source, the results of hotting-up the engine from Mark 1 to racing standard produced 230hp and 2400rpm for take-off. I have looked at one of the original engines, unmodified since its use in the MacRobertson race, and this has a stamp on the data plate quoting an output of 223hp.

Although the Comet project alone involved substantially more than just designing and building a conventional aeroplane, for by the standards of the day it was far from conventional, yet another aspect needed attention. There was no point in producing a potential race winner if no one was prepared to own and enter the new machine. Overall development and production costs were forecast to be about £50,000 and there was no question of a large-scale manufacturing run on which these expenses could be absorbed. Equally, though, no prospective purchaser would be prepared to pay an excessive price for a machine designed exclusively for the race and which might offer little subsequent value as a workaday aeroplane. The only solution with any prospect of commercial success was to look further than the Comet as an entity in itself: to subsidise the type heavily in long-term anticipation of achievement and prestige, with the possibility of subsequent benefits in orders for de Havilland designs generally. This arrangement formed the basis on which the

An early view of E-1 clearly showing the American controllable pitch propellers that were used for the first eleven flights (seven tests). The labels on the blades state 'Hamilton Standard Propeller Company'.

DH88 saw daylight, with a selling price to those who placed orders before 28th February of only £5,000 apiece and with guaranteed delivery in September.

Most of the construction work took place at the company's works at Stag Lane, Edgware, where DH82 Tiger Moths, DH83 Fox Moths and DH84 Dragons were keeping the production facilities at full stretch, but many of the associated tasks took de Havilland people to other places. The airframe and engine work, although demanding excessive efforts and often round-the-clock activity, could be carried out within the organisation, but the propellers, in particular, caused much searching in both heart and mind. Messrs F. T. Hearle (the General Manager) and A. E. Hagg visited the United States and before long they were heading for the Hamilton Works at Hartford in Connecticut. This company had variable pitch units already in regular use on the Douglas DC-2 and other airliners and the differences in performance between the machines with the modern propellers and those with the earlier fixed-pitch blades were very noticeable to the de Havilland men during their travels across America. They were impressed by what they saw, not the least by the technical efficiency of the Hamilton set-up. However, as always, there were difficulties. These propellers were in use on big radial engines and they were of a large diameter that would be quite unworkable on the small Gipsy Six; if they were cropped to size, the blade forms would not be fully efficient. Also, the pitch-change was carried out hydraulically, requiring special pumps and extra design time and adding more weight to an aeroplane that would need to shed every pound of non-essential equipment to trade-off against the amount of fuel that could be carried. The loaded weight, although not announced publicly until much later, was expected to be about 5,250 lbs.

Firm orders were placed with Hamilton, but too many unresolved factors remained for guaranteed success in the few months available. The next move was to investigate the products of the French Ratier company, which had developed a more basic idea of a propeller with two pitch settings; this could be fixed in fine for take-off and initial climb and would change automatically to coarse for the cruise. Although clearly not as fuel-efficient as a unit with controllable blade settings between the pitch stops, this would enable the Comets to get away smartly at full load from small aerodromes and then cruise reasonably economically. Indeed this stand-by arrangement proved not only sensible, but essential, for in the end the French unit was used.

The Ratier propeller was a relatively crude device, but its simplicity removed the need for pumps, lines between the engines and cockpit or even pilots' controls. The blades were held in coarse pitch by a spring on a piston, while for the fine setting the strength of the spring was overcome by air pressure in a bag

The first DH Gipsy Six R to be installed in the first Comet

inflated by a bicycle pump. On this basis the propeller would be pumped into its take-off setting before the engine was started; it would remain in this condition for take-off and initial climb, but when the aircraft reached a pre-set figure (in the case of the Comet, 150mph) the airflow pushed back a disc mounted in front of the propeller hub and this released the air from the bag, which in turn caused the spring to turn the blades into coarse pitch. So far so good, but once this cruise setting was established, the pilot had no control over the situation and if he needed to go round again from a baulked landing approach he would be forced to do so with wrongly-set blade angles. Another problem that appeared subsequently was a consistent inability to set both propellers to change pitch at the same time and any pilot with experience of flying multi-engined aircraft with unsynchronised engines will appreciate the discomfort that this must have created.

Work continued on seven days a week and often into the night. A half-scale model of the wing was built to try the constructional method and subsequently tested to destruction to prove its potential strength. A DH85 Leopard Moth, G-ACHD, was fitted with the first Gipsy Six R engine and this was flying on test in May of 1934; by this time only four months remained before the promised delivery date of the Comets to the customers.

The first order, within a week of the announcement, was from Jim Mollison, who would fly his machine in the race with his wife Amy, who as Amy Johnson already had established her reputation as a long-distance record-breaker. The second, on 23rd February, was from A. O. Edwards, managing director of the Grosvenor House Hotel in London, who nominated C. W. A. Scott and T. Campbell Black as his pilots. Considerably later than the first two, the third and final Comet for the MacRobertson event was ordered and entered by Bernard Rubin; although initially the owner declared that he would fly in the race, he was taken ill, so Owen Cathcart-Jones and Ken Waller became destined for the hot seats.

After construction work and trial assembly at Stag Lane, the first Comet was moved by road in three loads of wing, fuselage and engines to the company's new aerodrome at Hatfield, which shortly was to become the main operating base. The move itself was not the easiest of jobs, for the wing was in one piece with a span of forty-four feet and particular care was needed to avoid the possibility of damage or jarring en route.

As the first-ever British aircraft to have variable-pitch propellers, retractable undercarriage and trailing-edge wing flaps, the Comet saw daylight as a complete aeroplane as late as the first week of the month in which delivery to the customers had been promised. Prior to its registration later as G-ACSP and only six weeks before the start of the race, the first completed aircraft (constructor's No. 1994) took to the air on 8th September bearing the identity E-1. This did not stand for experimental, as some people have stated, but was a letter allocated to de Havilland for aircraft on test and was used on several machines, including the Leopard Moth on which the Gipsy Six engines had been proved.

Throughout the Comet's history from first thoughts to the weeks leading to

Before the first flight: E-1, later to become G-ACSP, at Hatfield early in September 1934.

the race, the entire operation was covered in secrecy; even the first flight was carried out very early in the morning with hardly anyone present on the aerodrome at Hatfield to see the momentous occasion. As the possible swinging tendency could not be known fully in advance, Hubert Broad took off gently; then he flew a large circuit that took him out of view of the few spectators. When he reappeared, his wheels were up and people on the ground feared that he was about to land with his undercarriage retracted, so, to warn him, one of the team quickly removed the spare wheel from his car and held it in the air for the pilot to see. However, he was not intending to land; he had lost count of the number of turns that he had given to the hand-wheel. The basic undercarriage retraction system had nothing as sophisticated as indicator lights and Broad was making a low pass in the hope of being able to see in the shadow from the low sun whether his wheels were up or down! The car wheel signal provided proof of his own findings and after winding down the wheels he came round for another circuit and a perfect landing.

Much publicity has surrounded the Ratier propellers as these were fitted when the Comet emerged into public view, but the first test flights were carried out using the Hamilton controllable pitch units. The earliest photographs show clearly that the French propellers were not fitted, as the spinners are

Hubert Broad aboard E-1 on the first flight. Noticeable differences from the Comets at the time of the race include the pointed spinners, very low cockpit canopy, absence of a landing light and a shorter tailskid.

From the left, Geoffrey de Havilland, Hubert Broad, Arthur Hagg, Charles Walker and Frank Halford in serious discussion immediately after the first flight on 8th September 1934. (Business Press International).

C.A. Form 71.

Classification for Aerodrome Charges.

Landing.—Under **7000** lbs.

Housing.—Under **1276** sq. ft.

GREAT BRITAIN AND NORTHERN IRELAND.

AIR MINISTRY.

Certificate of Registration of Aircraft.

No. **5063**

Nationality and Registration Marks ... **G-ACSS**

1.	Type and Description	D.H. 88 "COMET" Two-Seater Monoplane.
2.	Name and Address of Constructor ...	The de Havilland Aircraft Company, Limited Stag Lane Aerodrome, Edgware, Middlesex.
3.	Constructor's Serial Number	1996
4.	Name of Owner	Artur Octavius EDWARDS.
5.	Address of Owner	Grosvenor House, Park Lane, W. 1.
6.	Nationality of Owner	British.
7.	Usual Station of Aircraft	Hatfield.

WHEREAS it has been declared that the above-mentioned aircraft is not registered in any other State, it is hereby certified that the said aircraft has been duly entered on the register of GREAT BRITAIN AND NORTHERN IRELAND on the **4th** day of **September**, 19**34**, in accordance with the Convention for the Regulation of Aerial Navigation, dated 13th October, 1919, Annex A, and with the Air Navigation Act, 1920, and the Orders in Council in force thereunder, and has been allocated the Nationality and Registration marks **G - A C S S** and is of BRITISH Nationality.

Given at London this **5th** day of **September**, 19**34**.

AIR MINISTRY

for Deputy Director of Civil Aviation.

Attention is drawn to the extracts from the Regulations overleaf.

Note.—No entries or endorsements may be made in the foregoing certificate except in the manner and by the persons authorised for that purpose by the Secretary of State.

pointed at the nose with no protruding discs. Following various problems, including tendencies for the engines to overheat, the Hamiltons were removed and the Ratier propellers were fitted from test No.8 (believed to be the 12th flight) and for all subsequent flying. On the first flight with the Ratiers, the left propeller changed from fine to coarse at an indicated airspeed of only 110mph, while the right hand unit remained in the take-off setting until the aircraft had accelerated through 160mph!

The flying characteristics of any new design need very thorough testing at various speeds and initially at a safe height, but lack of time prevented the Comet from undergoing the very extensive programme that one might have expected for a type embracing so many new features. The tapered wings with pointed tips were an essential feature of the designer's performance target, but these were known to generate pronounced wing-dropping tendencies both at and near to the stall, so care would be needed during the later stages of the approach and especially during the hold-off and landing itself. Hubert Broad confirmed that this expected tendency had proved itself in practice and reported also that the DH88 was very sensitive laterally, necessitating the fitting of steel springs at the bottom of the control column to provide artificial built-in feel; otherwise he reported that the aircraft flew well and before long he was able to announce that at least the speed element in the specification could be matched. Official tests provided a published capability of 235mph at 1000 feet and 225mph at 10,000 feet, although these figures vary slightly from those on the flight reports. Extended fuel and oil consumption checks

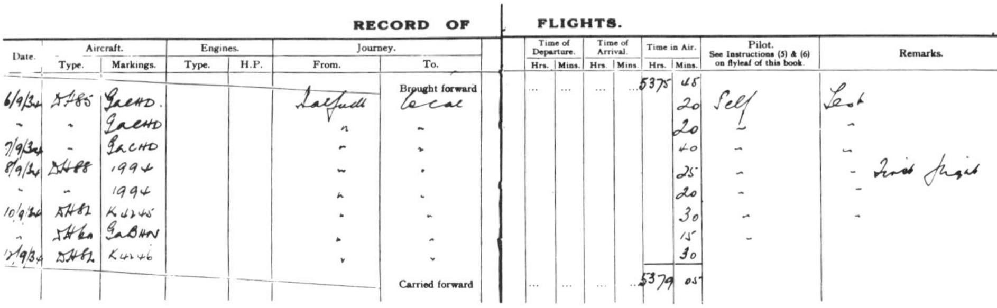

RECORD OF FLIGHTS.

Date.	Aircraft. Type.	Aircraft. Markings.	Engines. Type.	Engines. H.P.	Journey. From.	Journey. To.	Time of Departure. Hrs.	Time of Departure. Mins.	Time of Arrival. Hrs.	Time of Arrival. Mins.	Time in Air. Hrs.	Time in Air. Mins.	Pilot. See Instructions (5) & (6) on flyleaf of this book.	Remarks.
						Brought forward	...	...	...	...	5375	45		
6/9/34	DH85	GACHD.			Hatfield	Local						20	Self	Test
-	-	GACHD			-	-						20	-	-
7/9/34	-	GACHD			-	-						40	-	-
8/9/34	DH88	1994			-	-						25	-	- First flight
-	-	1994			-	-						20	-	-
10/9/34	DH82	K4245			-	-						30	-	-
-	DH80	GABHN			-	-						15	-	
12/9/34	DH82	K4246			-	-						30		
						Carried forward	...	...	...	...	5379	05		

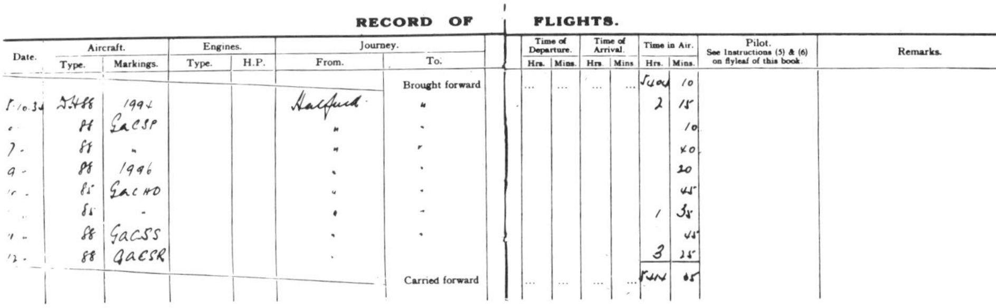

RECORD OF FLIGHTS.

Date.	Aircraft. Type.	Aircraft. Markings.	Engines. Type.	Engines. H.P.	Journey. From.	Journey. To.	Time of Departure. Hrs.	Time of Departure. Mins.	Time of Arrival. Hrs.	Time of Arrival. Mins.	Time in Air. Hrs.	Time in Air. Mins.	Pilot. See Instructions (5) & (6) on flyleaf of this book.	Remarks.
						Brought forward	...	...	...	...	5402	10		
5.10.34	DH88	1994			Hatfield	-					2	15		
6 -	88	GACSP			-	-						10		
7 -	88	-			-	-						40		
9 -	88	1996			-	-						20		
10 -	85	GACHD			-	-						45		
-	85	-			-	-					1	35		
11 -	88	GACSS			-	-						45		
12 -	88	GACSR			-						3	25		
						Carried forward	...	...	...	...	5412	05		

Extracts from Hubert Broad's personal flying logbook. The upper page shows the first Comet flight on 8th September 1934. This reveals that No. 1994, shown from 7th October as G-ACSP, was the first to fly. The second page shows the first flights by G-ACSS and G-ACSR, but see the postscript.

presented greater difficulties with such a tight time schedule. Early modifications included raising the height of the windscreen and cabin top by one and a half inches to improve the very poor forward view and lengthening the tailskid. Extracts from some of the flight test reports and a set of 'General Instructions to Pilots', published on 9th October 1934, appear as appendices 6 and 5 respectively.

The name of de Havilland was held high with the three Comets appearing in rapid succession. Within three days of each other G-ACSP was registered to Bernard Rubin, G-ACSR in Jim Mollison's name and G-ACSS was registered to A. O. Edwards. True to its word, the company had completed and flown all three machines in readiness for hand-over to their owners in time for the race.

The Comet's subsequent success story comes to light and life in the pages that follow, but this may be the stage at which to stress the speed with which everything had been achieved. The aircraft logbook for G-ACSS (the second to fly, on 9th October) reveals that the maker's testing amounted to only four short flights, totalling only one hour and twenty minutes in the air, while at the starting line, including C. W. A. Scott's time familiarising himself with the new aeroplane, 'SS had flown only three hours and thirty minutes since roll-out. By the time of arrival at Mildenhall, Scott had completed only five landings in the Comet and Black had landed twice.

The Comets were ready and, within days, the race would be on.

E-1, showing clearly the Comet's remarkably lean and clean lines. Was ***any*** *aeroplane aerodynamically smoother than this?.* (The Aeroplane).

Chapter 3
Mildenhall

Only three months before the race was due to begin, a starting point had not been firmly established. At one stage the rules would allow competitors to take off from the choice of four different aerodromes and fly over an agreed check point; only after several months of search and subsequent investigation was the newly-constructed aerodrome at Mildenhall, near Bury St. Edmunds in Suffolk, selected as the apparent favourite. However, when arrangements were almost finalised, some factions within the Royal Aero Club expressed their dissatisfaction about the remoteness and lack of accommodation; after all, they said, it was yet to become a usable Royal Air Force station and many of the buildings were only partially completed.

Several people proposed that Hatfield should be chosen, but as late as 6th September Mildenhall won the day. If offered a clear take-off run of 1400 yards and as the Service had not yet moved in the new hangars were unoccupied. The Air Ministry gave approval subject to certain financial arrangements for its use.

The Race Committee arrived at Mildenhall on Sunday 14th October. Already crowds had assembled to see the competing aircraft and already the police appeared to have lost control of the security arrangements. The Clerk of the Course, Commander Goodfellow, declared that he could not continue with his work unless reinforcements were brought in, while more adequate plans were needed to keep the people away from the active areas, in which many crews and engineers were making eleventh hour adjustments to their aircraft. Although there were six days to go before the flag would be lowered for the first machine to start on the epic flight to Australia, allegedly 3000 people had broken through the enclosure barriers and at least four competitors had complained of damage to their aeroplanes. Other arrangements were rudimentary, to express the position politely. There was no windsock, so at first a potato sack was hung from the pole, but this was both inefficient and inconspicuous, to be replaced by a pair of pillowcases hurriedly sewn together. At some stages a bonfire was used. Even such essentials to safety as Very pistols and signalling lamps were absent and the only available illumination for night work was provided by a pair of battery-operated arc lamps.

Preparing the aerodrome at Mildenhall before the arrival of the first aircraft.
(Business Press International).

Monday was the day by which all competing aircraft were required to arrive and by late afternoon twelve machines were present. Two others were reported to be on their way across the Atlantic. Although the three Comets had been completed and flown in time, problems of pilot familiarisation and weather created several setbacks. The first to arrive was Bernard Rubin's G-ACSR, which landed in a wind reported to be gusting to 40mph, so not surprisingly it suffered a minor wing-drop during the landing. Shortly afterwards Messrs. Scott and Black arrived safely in G-ACSS. The last Comet, G-ACSP, flown by the Mollisons, suffered from wind problems and Jim needed to go round again twice with the added difficulty of the propellers being set irreversibly in coarse pitch. By this time Bernard Rubin had declared that he was unable to fly in the race and he had arranged for Owen Cathcart-Jones to do so on his behalf.

Throughout this time all the available de Havilland expertise was firmly established at Mildenhall and giving all the required attention to the Comets' airframes, engines and equipment. There were plenty of problems, including a damaged undercarriage on G-ACSR which occurred only about 36 hours before the race was to begin, due to a landing with legs not fully locked down; this must have been a very easy and excusable error with the original winding system and there were several similar incidents during the type's subsequent career. Although the undercarriage and related airframe damage could be repaired, the propeller blades were bent and there were neither spares nor an approved straightening scheme.

Many people assumed that 'CSR would be a non-starter, for although the Ratier blades were taken to and re-bent at Fairey's works, neither the company's chief inspector nor the Aeronautical Inspection Department would sign that the repair work followed approved practices. However, a last minute brain-wave brought the French Ratier engineer into the picture and he was willing to sign the log book and certify the repairs. Fortunately his signatures were accepted by the other parties concerned.

This was not all. A last-day fuel consumption check on G-ACSS indicated that unless wind conditions were fully favourable it might not manage to complete the long lap to Baghdad without refuelling. To compensate, the carburettor jets were changed and with insufficient time for further flight tests, engine runs were carried out in the dark to check the mixture strength by the colour of the exhaust flames. Yet more difficulties were to follow and on the very last night before the race, new, larger oil tanks were fitted to all three Comets.

Weight was yet another problem. The Certificate of Airworthiness for each aircraft type clearly states the maximum limit and by this time the established permissible fully-loaded figure for the DH 88 was 5500 lbs. However, with tanks full and all aboard 'SP, 'SR and 'SS all tipped the scales at weights above this. Nothing could sensibly be removed, for the only articles not essential to flight were the required items of safety equipment to be carried for use following emergency landings. However, surprisingly, the difficulty was eased with the discovery that suitable fuel was available with a lower-than-standard specific gravity and this brought the Comets down to weights *nearly* within their legitimate limits. Then Harold Young, the Resident Technical Officer for the Air Ministry, authorised a last-minute increase of 50lbs on the Comet's maximum permitted weight and the problem was solved.

By this time considerable effort had been made in the production of printed material, including instructions to control and checking points, pilots' handbooks and official brochures. The rules laid down that each point would have a central control office at which a senior controller would be on duty to

Mildenhall Aerodrome a few weeks before the start of the MacRobertson race to Australia. (Business Press International).

There was no windsock at Mildenhall, so this bonfire served as a substitute.
(Business Press International).

inspect and sign each pilot's log-book and certify times of arrival and departure. Also at each point a technical officer would carry out all the inspections required by the rules of the race, while a third official would liaise between competitors, fuel companies, customs and other essential contacts. Among his other duties the controller would check on the availability of any necessary facilities for night landings, that obstructions were correctly marked, that first aid and fire/rescue facilities were laid on and ensure generally that everything was in good operational order for the nature of the flights to be carried out.

The written orders clearly explained the duties of each technical officer. He would be required to check that the pilot named on the entry was the pilot on board, that other occupants were the correct people with no crew changes en route, that the required rations (for three days) and safety equipment had not been shed in the interests of light weight and improved performance; also that the fuel load and total weight of the aircraft were as stated.

A practical publication known as the Guiding Brochure and Route Handbook was issued to each competitor; this provided all the relevant information to ensure that everyone knew what the requirements would be and how they were to be checked at points along the way. It explained also that all timings would be adjusted at Singapore to enable the finish at Melbourne to be realistic; for example if the last aircraft should leave Mildenhall forty minutes after the first departure, the latter would be held at Singapore for that time. By this arrangement, despite the differences in take-off times, in theory all the aircraft should finish at the same time, with the result that (subject to no disqualification) the first across the finishing line at Flemington racecourse really would be the winner. In practice this degree of accuracy proved to be unnecessary, but such precision of planning would have avoided some unpleasant situations that could have arisen.

Because the main buildings of the new Royal Air Force station at Mildenhall had not been completed, the place was a mass of marquees serving as offices, restaurants, bars and stores with only one incoming telephone line available to serve all requirements! Although the Royal Aero Club had laid down very comprehensive rules and little if any scope was left for the competitor with an imaginative mind, the atmosphere grew increasingly tense with queries, arguments, technical problems and requests for weight concessions adding to the tempo of the time. On the day immediately before the race, King George V and Queen Mary visited the site and both expressed their surprise at the apparent muddle everywhere. The King disliked aeroplanes and this was his first appearance at any civil aviation meeting, but to the Queen's delight he was persuaded to climb aboard and inspect a machine from within: not, unfortunately, one of the Comets, but the American Boeing 247D airliner entered by Roscoe Turner.

In the midst of what was generally reported as uncontrolled human chaos (with the noted exception of Alan Goodfellow, the Clerk of the Course, who was remarkable for remaining calm and good-natured throughout until he 'went under' with a temperature of 101° on the last day before the race) this message for all competitors arrived from Sir Macpherson Robertson:

> 'The magnificent response by leading airmen of the world, who have so generously come forward to make this event the greatest race in the history of aviation, is very gratifying to the air race authorities in Australia. The full growth of aviation must be world-wide and the international support that the race has evoked since its inception convinces me that such a contest must help to broaden the mutual basis for friendly exchange of services and understanding between nations, quite apart from its great value in quickening the airmindedness of their people. I but express the feelings of all Australians in wishing you a safe and speedy flight to Melbourne, where a warm and hearty welcome awaits you.'

The start of the race was very close. Throughout the final night last minute work was carried out on nearly all the competing aircraft and the three Comets were receiving more attention than most. Whilst engineers buttoned-up the aeroplanes, pilots dossed down in the backs of cars, in make-shift dormitories and wherever they could lay their heads for a few hours of horizontal restlessness; but the Comets and their crews would be (and were) ready. By 4am the participants were up and about; tea, coffee, toast, eggs and bacon were laid on at this unsavoury hour, but the time of morning and the pre-flight tensions combined to cause very few to sit and linger long over their meals.

Aircraft were everywhere, but here we must concern ourselves with our prime subjects — the three Comets:

Bearing racing number 19 and painted green, but with no name, was G-ACSR owned by Bernard Rubin, a rich Australian whose family owned

G-ACSP, 'Black Magic', the Comet flown in the race by Jim and Amy Mollison, arriving at Mildenhall.

Bernard Rubin's Comet G-ACSR at Mildenhall. The pilots were Ken Waller and Owen Cathcart-Jones.

several sheep stations. The pilots for the race were Owen Cathcart-Jones, who had served in the Royal Navy from 1922 until 1930 and who in 1931 had been co-pilot on a successful flight to the Cape in a Lockheed Vega; and Ken Waller, an instructor at Lympne in Kent who had taught Mr. Rubin to fly and had flown with him in April 1934 on a flight from Australia to England. In post-war years he became well known as experimental test pilot for Miles Aircraft. Reports of the time indicate that these two were the calmest and most organised of the three Comet crews.

G-ACSS, bright red, owned by A. O. Edwards and named *Grosvenor House*, carried the racing number 34. The first pilot was Charles William Anderson Scott, who at the time held the England-Australia record, which he had achieved in October 1932 by reaching Darwin in 8 days, 20 hours and 47 minutes. His father was Charles Kennedy Scott, who was a well known musician. C. W. A. (never known by his Christian names) flew in the RAF from 1923 until 1929, but following several courts martial for low flying and other unauthorised exploits in the air, he left, later to fly as a pilot with Qantas. His cockpit companion was Tom Campbell Black, who had learned to fly in 1917 with the Royal Naval Air Service and later served in the RAF. He formed Wilson Airways, the first aircraft operating company in East Africa, and in 1931 he flew the Prince of Wales on several occasions during a hunting expedition. Scott and Black had met in the bar of the Royal Aero Club and had agreed not to compete against each other in the MacRobertson Race. How they came to be crewed together is described best in Scott's own words:

> 'Strange is it not, how, knowing a person by reputation alone, one seems to know him intimately without having yet met or even seen him? Believe it or not, that was the case with Tom Black and myself until we met at a cocktail party in London last October. We got on famously from the moment somebody introduced us and we said our "How d'ye do's". There and then I decided to ask Tom to be my co-pilot in the event of my ability to fix up a machine for the job. In the Aero Club next morning I said to Black: "What about teaming up with me in the MacRobertson Air Race?"
>
> Tom did not stop to ponder.
>
> "Yes," he answered straight out. "Good idea, Scotty. I'm with you—all the way."
>
> Then over two cups of coffee we came to a solemn agreement that in no circumstances would either of us take a hand in the race unless we flew together. That pact cut out the possibility of having to fly with young and enthusiastic pilots of financial strength and capacity for adventure, but with none of the skill or the experience so essential in a partner in a race like the one then in prospect. Oddly enough, throughout the next few months we both received plenty of offers of this description.
>
> January came, and the outlook seemed pretty hopeless. We were both terribly keen to compete, but the financial demands were far beyond our joint resources.
>
> One day a certain gentleman offered us a "mount" in the race, on

condition that we shared with him the prize money and other profits if we won. The arrangements had been carried as far as a dinner to celebrate the signing of the contract when the gentleman came to inform us that "unfortunately the trustees could not see their way clear to provide the cash for such a *crazy adventure.*" Black and I were crushed with disappointment. The prospects of taking part in the race to Melbourne had receded still further.

That week-end I went out of London. Returning on the Tuesday morning the telephone bell rang just as I entered the house.

"That you, Scott?" asked a voice at the other end.

It was an official of the de Havilland Company. They knew already of our solemn agreement to fly as a team, and wanted to see Black and myself at Stag Lane immediately. Tom Black was in a nursing home at the time, so I went off to interview de Havillands alone.

Within an hour I was dashing back to town again to meet Mr. A. O. Edwards at the Grosvenor House Hotel, of which he was Chairman and Managing Director. In a brief business-like interview Mr. Edwards made it clear that he was determined to enter a machine, and had instructed de Havillands to build it. In twenty minutes the whole thing was settled, signed and sealed. Tom Black and I were booked for one of the Comets.

Early in February we had a look at the machine on paper, and both of us realised that if the real thing could be produced in the workshop, then we had the right aeroplane for the job.

C. W. A. Scott (left) and Tom Campbell Black, with G-ACSS 'Grosvenor House', on arrival at Mildenhall.

G-ACSS, 'Grosvenor House', at Mildenhall before the start of the race.

Delivery was promised on the first of this October, but by mid-July we were reconciled to the knowledge that we should be lucky to get it on the appointed day.

Throughout September we waited daily for news of the Comet, and daily we were disappointed. The first week of October passed, and still there was no aeroplane.'

G-ACSP was the mount of the Mollisons. Painted black and gold with the name *Black Magic*, it bore the racing number 63. In 1931 Jim Mollison had completed a record-breaking flight between England and Australia and it is to his credit that the race committee decided to use his route for the 1934 event. Like Scott, he had been viewed with some disfavour by the authorities when he served in the RAF and he became known as a publicity-seeking playboy. Despite this, his flying achievements earned him permanent fame. In 1932 he had married his co-pilot for the MacRobertson race, who as Amy Johnson already had achieved her own claim to fame, initially by her solo flight to Australia in 1930 in the Gipsy Moth G-AAAH which today is on view at the Science Museum. Later both Jim and Amy served in the wartime ferrying organisation, the Air Transport Auxiliary, in which Amy lost her life.

Although the Comets form the main subject matter for this story, they cannot be seen in isolation from the remainder of the field. The MacRobertson Air Race was to be a highly competitive event and would not be a walk-over for anyone. It had attracted the eyes and ears of the world and had drawn entries from an equally wide range of sources. Some had performances in both speed and range to be serious contenders for prizes in both the speed and handicap sections, while others of lower performance could compete only for the latter; but although they were to be run concurrently, in many ways they were to be treated as separate challenges and those who were eligible to do so set their sights on successes in both.

By late July the list of entries had been released by the Royal Aero Club. These comprised 18 from the United States, 14 from the United Kingdom, 8 from Australia, 6 from France, 5 from Holland, 4 from New Zealand and 2 from Italy, while Algeria, Denmark, Eire, Germany, India and Sweden produced one apiece.

However, for one reason or another, many of the 63 aircraft that initially had entered the event had been withdrawn, so we have no need to concern ourselves with those that would fail to be starters, but among the array of assembled aeroplanes were several capable of great achievement. Possibly the most notable among these was the Douglas DC-2 entered by KLM — Royal Dutch Airlines. Predecessor of the even more successful and world renowned DC-3/Dakota, the DC-2 was a cantilever low wing monoplane with a retracting undercarriage and flaps; also it had the benefits of the Hamilton controllable pitch propellers that de Havilland had hoped to use on the Comet, but which had proved to require too much development in the few months available to suit the small Gipsy Six engines. Unlike the Comet with its narrow cramped cockpit that just accommodated two people, the DC-2 had been designed for regular airline service with levels of crew comfort that were rare at that time. Also, there was more than ample leg-stretching and relaxing space in the large passenger cabin that could accommodate fourteen people.

Not only was the DC-2 a good aeroplane, but without doubt KLM was the pioneer among the world's airlines and in chapter 1 I have made brief reference to its aims and ambitions. Its past achievements had been handicapped only because there had been no suitable aircraft with which to fly the international routes. Now, with this new machine that could establish an entirely new era in the business of transporting passengers and mail, KLM could see daylight ahead. Although in itself the publicity to be gained from success in the race would be extremely valuable in promoting the company and its capabilities, the race would provide other benefits in the form of experience on a long haul route operating to an extremely tight schedule. Although no subsequent regular service could hope to run to a timetable comparable with the possible race timings, the experience to be gained en route in the form of aerodrome capabilities, refuelling and technical turn-round times and weather penetration would serve as valuable guide-lines in future planning. So for KLM the race was no gimmick. The notably calm crew of Messrs Parmentier and Moll were very experienced, fully prepared and were backed by an organisation that exhuded nothing but professionalism. The event would serve as a possible pointer for the airline's future operations.

Despite the length of the route, more than half the entries were for aircraft with only one engine, but among the multi-engined machines the Boeing 247D transport entered by Colonel Roscoe Turner (a film stunt man and former lion tamer) warrants special mention. Although officially issued with the racing number 5, its intrepid operator insisted on displaying the figure 57, for his machine was heavily sponsored and, as he explained dogmatically, Mr. Heinz had paid a lot of money for it! On more technical grounds, though, the Boeing was no baby. Powered by a pair of 550 hp Pratt and Whitney Wasp nine-cylinder radials and with a span of 74 feet, it was only just down the power and size scale from the DC-2, which ran to 715 hp aside and 85 feet respectively.

Already in regular service in America with United Airlines, clearly the Boeing 247D was a machine to be watched. Roscoe Turner had experienced many problems in reaching the stage of the race start, firstly in arguments with the Royal Aero Club whose officials had refused for some time to accept the Boeing as a late substitute for a Douglas; later with Customs, for due to heavy seas the ship carrying his aircraft had been diverted from Southampton to Le Havre, where the French seized it and demanded £4000 in duty; and finally at the last minute weigh-in for the race, when his aeroplane pushed the scales to 300 lbs beyond the maximum permitted limit. After exhaustive searches for the cause, the reason was quite simple: he had calculated in American gallons at 6.2 lbs, while Imperial gallons pumped in were recorded as a nominal 7.5 lbs,

'Grosvenor House' being weighed at Mildenhall.

A general view of activities at Mildenhall shortly before the race.

although the fuel supplied at Mildenhall proved to weigh 7.7 lbs per gallon. Eventually Turner was forced to shed some petrol and lose some of his machine's planned range, but as with the Comets the difficulty was overcome partially by finding and using a fuel of lower specific gravity. All these setbacks made an already ambitious entrant even more determined to succeed.

At the other end of the weight, power and size scale were several types that would need to fight their ways around the world against enormous odds. Notable among these was a diminutive Miles Hawk Major entered by the Manawatu Aero Club of New Zealand, with a crew comprising M. C. MacGregor who had served under Billy Bishop, V.C., in the first world war, subsequently with the New Zealand Air Force and who now was the Manawatu Club's Chief Flying Instructor; and Henry (Johnnie) Walker, a Scot whose family had emigrated when he was a child. The Hawk was one of the many light machines that could compete only in the handicap section of the race.

Although the aeroplanes and the people associated with them, whether crew, engineers or on the organising side, formed the centre of everything concerned with the great event that was about to begin, the outside world had not allowed the occasion to pass unheeded. Following much advance Press publicity, the race seemed to appeal to high society of the day as much as to those with genuine interests in sporting aviation; many hundreds of people appeared in the early hours of Saturday 20th October still garbed in dinner jackets, or evening dresses and flimsy shoes, from their previous evening's more normal social activities. However, to put the picture in the words of one who was present I quote from the book by the famous actress Florence Desmond, who later was to marry Tom Campbell Black:

'The day before the race I motored from London to Mildenhall. The roads were packed with lines of traffic. Suddenly the whole world seemed to be interested in the race. Tom told me to meet him in the club tent. The place was a hive of excitement. Most of the pilots were there, and there was a tenseness in the atmosphere I don't ever remember experiencing before. To me the whole thing was rather frightening. These keen pilots, the best the world had to offer, were about to compete in a race which was to make flying history. Both Tom and Charles Scott had been training for weeks for the race. They were just about as fit as any two men could be. I got there in time for lunch. The large refreshment tent was packed with relatives of the pilots, friends, Press men, aircraft designers, the representatives of the firms supplying the fuel, and the backers. Champagne corks were popping, but neither Tom nor Charles took a drink, as they had been on the 'wagon' for many weeks.'

The organisers hoped to control the number of visitors by charging five shillings per person, but altogether about 60,000 people came to or near the aerodrome to see what was going on. In the early hours it was cold, the wind was blowing at an alleged 30 mph and for part of the time rain was falling. The approach roads were crammed beyond capacity and cars were abandoned for miles around, many prospective spectators tramping their ways to the field through ploughed furrows.

The Prince of Wales at Mildenhall with C. W. A. Scott, Tom Campbell Black and 'Grosvenor House'.

Here, perhaps, these words from C. W. A. Scott's own book may help to give another view of the Mildenhall story:

> 'In the final week I managed five landings. Tom did two. And, in case I forget to tell you, that week at Mildenhall, as an experience, was worse than the flight itself. It was filled with worries and anxieties, with attention to the regulations and official inspections, all leading, of course, to the night before the start.
>
> And what a night!'

By about five o'clock the aircraft were on the move. Some were taxied and others were manhandled, but altogether the frenzy of essential activity was hampered once again by the public bursting its legitimate bounds. Ten thousand or more people were said to have surged onto the active area of the aerodrome and the loudspeakers changed their tune, from providing weather information to crews, to calls for police and firemen to mobilise themselves into action. Motor cycles, fire tenders and other official vehicles proved to be very effective in restoring order and within a few minutes all was reasonably calm again.

The aircraft were ready (some only just), the crews were tense and the crowds could barely wait. By then the weather was tolerable and the strong wind may have provided some encouragement for those pilots whose machines' performances were being stretched to their limits for fully-laden take-offs.

This was Mildenhall at the end of a busy week of pre-race preparation.The zero hour was only a few minutes away.

Chapter 4
The Great Race

The early morning of Saturday 20th October was cold and damp, with the added burden for spectators of a wind blowing at nearly thirty miles an hour. Despite this, the crowds were insatiable in their demands to see the start of what has been described as the world's greatest-ever sporting aviation event. Strangely, although all concerned on the organisational and operational sides of the race were tensely occupied in completing all the last-minute arrangements and whilst the feverish activity on the ground created an atmosphere that flowed over to those who had come just to stand and watch, there would be relatively little real visual excitement at this stage: just a succession of aircraft staggering off at full loads and taking unusually long runs to get away from the ground, followed by relatively gentle turns onto course for the first leg before each disappeared into the cloud-laden dawn sky.

Whether the endless pre-race Press publicity (not matched before or in the fifty years since) had stirred people into expecting dramatic and dashing sights or whether the socialites attended in their thousands just because they were expected to be there we may never know, but despite the physical discomforts they were on and around Mildenhall aerodrome to see what the MacRobertson Air Race was to be all about.

The official Royal Aero Club starter was A. G. Reynolds and it was his task to ensure that all the competitors were launched in the correct order; also that they were despatched at the time intervals related to their calculated handicaps previously worked out by the well-known pair F. Rowarth and W. Dancy. However, Mr. Reynolds relinquished his flag temporarily for the first aircraft to be waved off at 6.30 a.m. by Sir Alfred Bower, the acting Lord Mayor of London. That first aeroplane away was the Comet G-ACSP, *Black Magic,* with Jim Mollison at the helm and his wife Amy tackling the daunting task of navigation. I have spoken to several people who were present for this and they have confirmed their feelings about the anti-climax from a spectator's viewpoint; collectively their clearest memories are the relatively slow initial acceleration and the seemingly endless take-off runs of aircraft that were loaded to their limits. One wonders whether some of the runs would have been endless, or how and where they might have ended, if the day had dawned with a flat calm!

To the casual onlooker there may have been relatively little to cause genuine visual sparkle, but for those in the cockpits the next few days would be the most harassing and skill-testing times that most had experienced at any points in their lives. Problems of long-range navigation with no artificial aids as we know them today; difficulties of flight in bad weather, often over barren and mountainous ground and without the full set of comprehensive flight instruments that we accept as standard; cramped cockpits with no personal manoeuvring space for hours and hours on end; and the need for reasonable accuracy in both track and time-keeping, not only to achieve success in the contest but in order to reach the destination at all, with fuel and maximum range offering only a small safety factor over the length of the leg. 2,530 miles lay between the point of take-off and Baghdad, which was to be the first target for some of the competitors in the speed section of the race. The feelings within the cockpits must have varied from one pilot to another, but even the calmest of competitors must have faced fear with the thought of all the difficulties that lay in wait between the two points, with several similar and in some cases more hazardous legs to follow.

The second aircraft to be flagged away, after the Mollisons, was the U.S. registered Boeing 247D airliner NR 257Y with Colonel Roscoe Turner as skipper and Clyde Pangborn as co-pilot. Turner was a great showman with a circus background, but despite wearing a rather garish uniform of his own design and indulging in many pre-flight gesticulations for all to admire, only the starter's flag and the rather unusually rapid opening of the throttles on the airliner's two Pratt and Witney Wasp nine-cylinder radial engines made this appear to be different from a routine take-off on a regular scheduled passenger or mail service.

The third aircraft to be unleashed differed from its predecessors in providing some unintentional entertainment for the spectators and some concern among those who cared. Only just ready after the repairs to its undercarriage and propellers, the unnamed green and white Comet G-ACSR, (with Ken Waller and Owen Cathcart-Jones destined to share the poles, pedals and throttles

Competition: Roscoe Turner's Boeing 247D which was the third aircraft to finish, ahead of Comet G-ACSR. Note the 'Heinz' number on the nose.

along the route) had mixture problems, with white flames emitting from one exhaust during the take-off run. The attempt was aborted, but the aircraft was stopped just inside the aerodrome boundary and taxied back to the start point for a second, uneventful and successful departure. Now two heavily-laden DH 88s were on their ways towards Baghdad.

Since the original entry list had been issued showing sixty-four aircraft, the number had dwindled drastically as many aircraft and/or crews clearly would be unable to compete. Although the eventual number at the start-line was down to twenty aeroplanes, mainly they were the more likely contenders and the relatively small size of the final field had little effect on the seriousness of the competition to be faced. Shortly after G-ACSR, the Dutch Pander S4 PH-OST and the British Airspeed Viceroy G-ACMU were launched into the fray.

Although the Comet was the only new aircraft designed and built especially for the race, the Viceroy deserves mention here. A development of the Airspeed Envoy (a twin-engined feeder-liner which had seating for six to eight passengers), the Viceroy was all British and had been produced solely for the race, to the order of Neville Stack, who flew it in company with S. L. Turner. Only one was built. It had no windows in what would normally be the passenger cabin and much of the left side of this area was occupied with extra tankage providing a total of 270 gallons. The other side contained a folding canvas bed and there was a lavatory at the back. Instead of the original Envoy's Wolseley radials of 185hp., 'CMU boasted the first-ever pair of Armstrong-Siddeley Cheetah VIs producing 290 hp. apiece. The result was a top speed of 210 mph, giving a bonus of about 40 mph over the performance of its progenitor. Unfortunately, however, the Viceroy suffered from a series of troublesome faults throughout the flight and retired hurt at Athens.

The sixth machine in the order of take-off was the third and last of the Comets to be launched: the red and white G-ACSS, *Grosvenor House,* with C.W.A. Scott and Tom Campbell Black aboard. The take-off, although uncomfortably long to the spectators, was uneventful and soon 'CSS was in pursuit of 'CSP and 'CSR heading for Baghdad, but here Scott's own feelings must be more descriptive than any bare historical facts:

> 'Through the dark hours of the race itself I was never so uneasy as in those moments of inactivity in the cockpit of the Comet, listening to the engines turning over, and waiting for the flag to fall. It was impossible then to appreciate the enormity of the venture.
>
> I just had time for a quick smile round, as with the two purring engines roaring into life we were off on the Baghdad trail.
>
> I had never taken the Comet up on full load before, but after a hundred yards I was perfectly and blissfully sure that in her we had not only a thing of beauty, but infinite power to play with.
>
> Low mists drifted beneath us as Mildenhall passed from sight, and once over the English coast-line we were swallowed in the haze of the North Sea. Almost subconsciously I seemed to understand that we were flying to Australia, but there was no evidence of the fact. Overwhelmed by the implications I turned round to Black and said, "I suppose we are

More competition: The Douglas DC-2 'Uiver' of KLM. This was the second aircraft to pass the finishing line at Flemington and throughout the race was a cause of concern to the crew of G-ACSS.

going to Australia?" Tom laughed. "I feel like that, too," he shouted back.'

Close upon the track of the last Comet to be launched was the most worrying contender of all: the Douglas DC-2 PH-AJU of KLM, named *Uiver* (meaning Stork). No organisation and no crew could have spent more time and effort in preparing for the race than that shown by the cool determination of the Dutch airline. Whilst the task would not be easy for anyone, KLM had the advantage of previous route experience over much of the way, a big aeroplane with plenty of space for the occupants and the scope for provision of support facilities at the check points that only a large company could muster. The two pilots, Messrs. K. D. Parmentier and J. J. Moll, exhuded a quiet calm professionalism throughout and probably were the only participants not to display those unpredictable variations in human behaviour that beset most people at times of tension.

The remainder of the participants followed at short intervals with the repeated lowerings of the starter's flag. Although many of the machines were relatively light types with chances in the handicap section only, at that time Britain produced most of the world's best light aircraft and some encouragement may be gained by knowing that nine of the thirteen to take-off after the DC-2 had English origins. The last person to be released from that intolerable wait at the starting line was C. T. Melrose in the Puss Moth VH-UQO.

Here a brief reference to the publicity surrounding the event may help to produce a picture of the way in which the world in general and Britain in particular was viewing the scene. Some newspapers devoted entire pages to progress reports and possibly no single event has attracted so much publicity. In the columns of informative detail were expressions of editorial opinion: among them the *Daily Telegraph* exclaimed it as 'the greatest air contest ever held', while true to form *The Times* provoked reasoned thought with 'if the aeroplanes live up to their promise there may be something like a revolution in the world's normal standards of time in relation to distance'. There were many more.

Soon progress reports were coming in. The KLM DC-2 was the first to arrive at the Rome (optional) checking point and was away again after a refuelling stop of only twenty minutes. The same machine was first to arrive and land at Athens, where the stop was reduced to nineteen minutes. Many of the machines were appearing at the optional check points, but here we must divert our thoughts from the activities of the rest of the field to concentrate on the performances of the aircraft with which we are primarily concerned, for just as the MacRobertson race was the reason for the DH 88 Comet's existence, equally the reason for this book is the DH 88 Comet.

After 12 hours and 40 minutes in the air, Jim and Amy Mollison in G-ACSP were the first to arrive at Baghdad and they were the only crew to cover the 2530 miles from Mildenhall to the first compulsory check point without the need to land en route for fuel. On this first long leg they had averaged 199.77 mph; they had flown for most of the way at about 10,000 feet, but for part of the flight had needed to climb to 16,000 feet to escape the weather and to keep the sun in view for navigation checks. Despite the length of time spent in the Comet's cramped cockpit the Mollisons took no rest, for they were on the ground for only an hour and thirty-five minutes, after which they were away again, with Karachi as their next destination.

Almost as soon as the Mollisons had departed, *Grosvenor House* appeared in the Baghdad circuit. Its arrival was one hour and fifty two minutes behind 'CSP, as Scott and Black had encountered several problems. Allegedly their first positive fix after leaving overcast Mildenhall had been the Danube and after an hour or so of reasonable conditions they faced further problems in heavy black thunderclouds over Turkey, with the prospect of not being able to find Baghdad if ever they should reach it. Worried about their whereabouts and their fuel—and no doubt remembering the problems about the aircraft's unproven range described in the previous chapter—the crew of 'CSS landed at the RAF aerodrome at Kirkuk. They wasted no time, however, for in less than half an hour *Grosvenor House* had uplifted sufficient fuel for the remainder of the leg (only 130 miles on a southerly heading) and was away.

At Baghdad, the turn-round was equally impressive. Despite the difficulties experienced on the first leg of the route, neither the crew of *Grosvenor House* nor the refuelling team wasted any time, for again in only thirty minutes the bright red twin was on its way towards Allahabad, 2300 miles away on a true track of about 130 degrees. Although most participants had planned to stop at the optional check point at Karachi, Scott and Black aimed from the start to go

جمعية الطيران العراقية

منهاج السباق ويعتبر بطاقة دخول

الى

الميناء الجوى ببغداد

يوم السبت فى ٢٠ تشرين الاول ١٩٣٤

بمناسبة السباق الجوي

بين انكلترا ــ اوستراليــا

القيمة **١٥٠** فلسا (لمنفعة جمعية الطيران العراقية)

IRAQ AEROPLANE ASSOCIATION

ADMISSION PROGRAMME

TO THE

BAGHDAD AIRPORT

ON SATURDAY, OCT. 20th, 1934

ON THE OCCASION OF

THE AIR RACE

ENGLAND AUSTRALIA.

Price : 150 Fils *(for the benefit of the Association's Funds)* all the way without landing; in fact, their ambition was realised, for, by contrast with their first leg to Baghdad, their path from Iraq and over Persia into north-west India was almost trouble-free and they landed at Allahabad at 9.18 a.m. (GMT). They were the first crew to arrive.

While Scott and Black were forging ahead with unexpected speed, life for the occupants of the other two Comets was proving to be less happy. The Mollisons had landed at Karachi after a total time of 22 hours and 13 minutes, thus establishing a record in the process, but after the next take-off the Comet's crude undercarriage revealed its limitations and *Black Magic's* wheels would not retract. 'CSP landed back for rectification, but after a delay of nearly sixteen hours a further set-back hit the Mollisons: immediately after the second take-off from Karachi, Jim realised that he had left his maps on the ground and needed to return to collect them. The final take-off (third time lucky?) was at 9.05 p.m. GMT and the machine was on its way towards Allahabad.

The crew of G-ACSR had suffered serious navigational problems over the Taurus Mountains and over the Black Sea. Map-reading was impossible and they were lost in total darkness. Low on fuel, they saw the lights of a village and luckily landed safely in the desert alongside. Very quickly the locals, who appeared to be Persians, were all round the aircraft, but they were not hostile. Communication was impossible and eventually both Cathcart-Jones and Waller decided that they must have overflown Baghdad and would need to return on a westerly heading. Although this involved covering some of the ground twice and would result in losing more time, they would be disqualified if they failed to report at the compulsory control point. They waited for first light and flew back to Baghdad, which they reached in little over thirty minutes but with virtually no fuel remaining in the tanks.

The unnamed 'CSR was on the ground for only forty-eight minutes before it was airborne on the next leg. However, the aircraft was soon showing signs of sickness, for long before reaching Bushire in the Persian Gulf, which is little over one-third of the way to Karachi, the starboard engine started smoking badly and was shut down, necessitating a return to Baghdad 'on one'. The offending Gipsy Six R was opened-up for inspection and after nearly six hours of intensive activity inside the cowlings, it was declared fit for flight again. 'CSR left Baghdad for the second time at 12.24 GMT, to arrive at Allahabad safely eighteen hours and eight minutes later, this including the refuelling stop at Karachi.

At this stage the Mollisons were in serious difficulties. The compass was giving erratic readings and any positive navigational fix was impossible. Totally lost, or as pilots tend to say nowadays, wholly uncertain of their position, Jim and Amy flew round in circles awaiting daylight and the hope of a fix. When they landed at dawn their tanks were nearly dry; they were at Jabalpur, about 175 miles south-west of their planned destination. Fortunately there was a bus garage in the immediate vicinity and the Comet was refilled with a low grade fuel that was entirely unsuited to the high-compression Gipsy Sixes that had been modified to racing standards. This

Action at Allahabad, with 'Grosvenor House' receiving replenishment.

must have been an early example of the problems caused by the wrong use of what now is called Mogas, which has been the subject of considerable controversy in light aviation circles over the past few years.

This marked the end of the Mollisons' participation in the race. Impatient to make-up for lost time, Jim flew at full throttle on this leg from Jabalpur to Allahabad, but the combination of already over-worked engines and low grade fuel had caused fluctuating manifold pressures and overheating on both sides, while one engine cut and picked-up intermittently several times. On arrival at Allahabad, the worse of the two engines was stripped and found to be in very poor shape; heads and pistons were badly burned and although both pilots hoped that repairs could be carried out, this proved to be an impossible task and the unfortunate result was close of play for *Black Magic* and the Mollisons. The stress also heralded the breakdown of their marriage.

Now there were two. Attention centred on Allahabad, where facilities were good. A beacon for night flying, (which at least one pilot claimed he could see from 70 miles away) a meteorological office, a control tower and powered fuel pumps were available. G-ACSS was on the ground for only 47 minutes and at 9.55 a.m. Scott and Black were forging ahead across the Bay of Bengal to Singapore, which lay 2210 miles ahead to the south-east. However, they were not wholly happy and Black, in particular, expressed concern at the way in which the engines had been running. Long legs, heat, virtually no rest and frequent turbulence all combined to tire the competing crews and the two aboard *Grosvenor House* were no exceptions. Scott, though, was reported to be relatively relaxed, although perhaps this paragraph from his book tells the truth:

'The weather report for the Allahabad-Singapore run was so bad that I kept it to myself. By that very act Tom must have guessed its nature, but asked no questions. We took off from Allahabad in trepidation and with a full load, on what we were prepared to believe must be the worst stage of the race.'

The reason for the very short stay at Allahabad was the knowledge that the well-organised KLM crew in the DC-2 were not very far behind. Scott had established that if he could be the first away from this holy Hindu city and so long as there were no technical delays, he would have a good chance of staying ahead for the rest of the race. Indeed, *Uiver* arrived in good order, with the crew still in neat airline uniforms. They, too, took only the minimum time needed for reporting, refuelling and eating. Parmentier calmly told the waiting Press that it had been one of the smoothest journeys in his experience, that there had been no incidents, that he would nurse his engines until reaching Singapore and then would open out for the final stage. It was all very professional.

The third aircraft to arrive at Allahabad created a different impression. The three-engined Dutch Pander S4 PH-OST had maintained excellent timing to arrive shortly after the DC-2, but it had behaved erratically at stages on the route and now, arriving in the dark, the crew of Asjes and Geysendorfer appeared unaware that the starboard wheel was not fully down and locked. Not surprisingly there were showers of sparks and the machine slewed round almost immediately after touching down. After three days of repairs, the Pander was in the air again on a local test, but allegedly it hit a beacon and caught fire in flight before crashing on the airfield. This was another machine now out of the running.

About three hours after the Mollisons had landed at Allahabad, G-ACSR arrived. This was at 8.32 a.m. GMT on Monday 22nd October. The crew achieved an almost record turn-round of little more than thirty minutes, but after take-off one propeller gave trouble with the jamming of the adjustment gear at the point when air pressure on the disc should cause the blades to change from fine to coarse pitch. Rectification took about five hours, after which Waller and Cathcart-Jones were airborne again en route for the check point at Rangoon on the coast of Burma.

The flight to Singapore appears to have been trouble-free, finishing with another rapid half-hour turn-round, but early on the following leg the starboard engine progressively lost oil pressure. Unless caused by a physical shortage of oil this is a sign of trouble within the works and is one of the most significant warnings available to a pilot; in fact many experienced aviators consider the oil pressure gauge to be the most important instrument in the cockpit. Although clearly unable to continue to their destination, with the sick engine throttled back the crew of 'CSR managed to fly on to the first check point, which was Batavia on the north-west coast of Java. Here, technical staff of KLM (which, we must remember, had its own entry to consider) found the problem and cured it, but by this time 'SCR was behind the hoped-for schedule and arrived at Darwin at 1.13 p.m. GMT on Wednesday 24th October.

Now we must return our attention to *Grosvenor House.* On the way to Singapore the Comet crew met some very rough weather; they had intended to

A posed refuelling scene at Allahabad, with (left to right) 'Dutch' Leonard of Mobil Oil, Rex Pearson, Chief Engineer, Ken Waller and Owen Cathcart-Jones.

take turns at flying throughout this leg but it was so turbulent that both pilots needed to hold the controls to keep the machine in any semblance of level flight. They flew through the night, but later when conditions calmed a little, Black in the back flew the aeroplane while the occupant of the front seat endeavoured to grab a spot of sleep. At daybreak through a gap in the clouds they recognised one of the Nicobar islands, but there would be many navigational problems ahead when the ground was out of view, so they reduced height to one thousand feet to remain below the worst of the weather. Otherwise they might have overshot the end of the Malayan peninsular. Eventually, however, Singapore appeared ahead and Scott needed to be woken to prepare for the landing, which, due to the Comet's poor view and lack of instrumentation in the co-pilot's position, needed to be carried out from the front cockpit. Both occupants were fatigued and only when Scott was committed to a landing did Black shout a warning that he was doing so downwind. Some reports say that the approach was crosswind. Possibly it was both, so I elaborate a little on this in Chapter 7. The arrival on the ground was hairy, to put it mildly, but Scott handled the situation and the aeroplane with commendable skill and no damage was caused. They were more than 7000 miles from England and had taken only forty hours to get there. The time was 10.31 p.m. GMT and on this turbulent stretch they had been in the air for twelve hours and thirty-six minutes.

Strangely, at this point, four of the key aircraft were on the ground at the same time, but all at different aerodromes. Scott and Black were at Singapore, Parmentier and Moll in the DC-2 were 1200 miles behind at Rangoon, Turner and Pangborn in the Boeing 247D a similar distance behind at Allahabad, with

Cathcart-Jones and Waller a further 900 miles behind at Karachi. Although 'CSS appeared to have a comfortable headway, each aeroplane was only a few hours flying time from the next and a simple technical delay could cause any such lead to be lost.

Scott and Black were on the ground at Singapore for an hour and eleven minutes before they left for Darwin, a distance of 2084 miles. Although shorter than the earlier legs the route was far from friendly, with up to 900 miles to fly over the inhospitable and shark-infested Timor Sea. Although they aimed to cover this stretch in one flight, for navigational purposes the crew had planned to overfly Koepang on Portuguese Timor Island as a time-check (it was available as one of the optional landing points if needed) before facing the long open water crossing.

This leg was far from easy for the Comet's crew. The worst part was at night, with low cloud over the sea, when again they let down to one thousand feet to remain clear of the weather. However, when tired, at low height at night and over seemingly endless water, this is a situation when all pilots pray that their engines will continue to behave well; but, alas, this was not to be.

Here I quote again from Scott's own report:

'I hate and loathe the Timor Sea. If anything happens to the motor there, then that's the end of pilot and machine.

The fine weather began to desert us, and at nightfall low clouds forced us down close to the sea. We were flying at about one thousand feet when suddenly one of the oil pressure gauges began to flicker. Tom immediately climbed for height in case things happened. Pressure dropped from 40 to zero, and Black throttled the engine back to prevent seizing. Here was a pretty predicament. Half-way across the Timor Sea, after struggling for two days through mad endeavours, trials and tribulations, were we to be cheated by a mechanical defect?

Not only that, but there was the terrifying prospect of helpless descent to the shark-infested waters below, and swift, certain death. In the pitch blackness we were still an hour from land.

Suddenly we sighted Bathurst Island just where Bathurst Island ought to be. By going a long way round one is able to follow a whole series of islands from Bathurst towards Darwin. With one engine disabled completely, that is what we did in the Comet. We hugged those islands tightly.'

The crew were aware that the other engine had been working excessively hard and were worried that the additional strain of using full power, needed to maintain height in asymmetric flight, might be too much for it for continue running smoothly. However, the remaining distance to go involved little more than a further thirty minutes 'on one' and the Comet arrived in this condition for a safe circuit and landing at Darwin, the nearest aerodrome to Australia's northern tip. Despite the small town's relative isolation, thousands of people were awaiting the first machine to arrive and the crowds broke their bounds to smother the Comet and its crew with excited congratulations.

Although Scott and Black were making excellent time, we must not overlook the efforts and achievements of the other competitors. The KLM DC-2 was continuing in persistent pursuit and this remained a primary incentive for the crew of *Grosvenor House* to keep on the move without unnecessary delay. The Dutch airliner was making very steady and consistent headway but was taking the entire route in far shorter stages than those used by the Comets. This may seem surprising, for clearly there would have been neither space nor weight problems in carrying extra tankage to enable the machine to complete longer legs, but KLM's commercially-orientated aim was to prove the practicality and reliability of air travel and not to indulge in gimmickry. Theirs was a standard DC-2 as in use in numbers in the United States and their attitude was far divorced from the idea of a specially designed and built racing aeroplane that might beat records, but which was of little use for anything else. Between Allahabad and Darwin the airliner made seven refuelling stops, with the longest time on the ground of 46 minutes and the shortest of only 22 minutes; everything was undertaken with clockwork efficiency. Only one stage behind the DC-2 were Turner and Pangborn in the Boeing 247D, so they, too, needed to be considered with respect.

Back to *Grosvenor House;* both pilots were understandably edgy in the knowledge that their engine problem might take a long time to rectify or might even be incurable in the time available. They had almost nightmarish visions of the Dutch airliner arriving at Darwin while their machine remained sick on the ground. At this stage Qantas engineers had the responsibility for servicing and they went through the usual sequence of removing components from the engine that had given so much trouble. Although two cylinders were very low on compression they found nothing in the way of mechanical failure and eventually the main problem was traced to a completely clogged oil filter. With this cleaned, with the engine reassembled and with new plugs, all of which had taken what seemed to the crew an interminable time but which in practice was completed within two hours, the engine was given a ground run. Clearly tired, it developed reasonable oil pressure and an acceptable output, but ran roughly. It was declared fit for the penultimate leg on a south-easterly heading to Charleville (1389 miles), where two new cylinder heads would be made available for fitting prior to the last lap into Melbourne.

On telephone advice from Frank Halford, the designer of the Gipsy Six (who had seen a placard in London stating that the leading Comet was grounded at Darwin with engine trouble) Scott opened-up very gently on take-off and stopped short of using full power, resulting in a very long take-off run; also he had on board only sufficient fuel for the length of the leg, with just a very small margin for navigational error or any other cause of delay. Both pilots were too tired to cope effectively and each took a turn to fly the machine for only a few minutes at a time. Before long Scott realised that he was lost, for although previously he had flown in the area for four years and had experience of the route on his earlier record run, he was unable to identify anything. Eventually, though, a railway appeared and he recognised this as the line to Charleville. Although well off track, at last he knew where he was and soon *Grosvenor House* was on the ground.

The first sentence spoken by the Comet crew was the question 'Where are the Dutchmen?' Almost at this very moment the DC-2 was touching down at Darwin; the gap was closing, for despite the airliner's frequent refuelling stops the large machine had behaved impeccably and had not suffered any technical delays. After completing the essential formalities Scott and Black were determined to take off again immediately for the final stretch to Melbourne. However, when they returned to the Comet, not only did they find the cowlings of the port engine off the aircraft, but the heads had been removed from the two offending cylinders and the new items that had been laid on by Qantas were about to be fitted.

As soon as the work had been completed and the Gipsy Six had been ground-run successfully, the crew of *Grosvenor House* took off again, but almost immediately the oil pressure trouble returned so there was only one move to be made: a landing back at Charleville. Although there are no recorded reports of the cause of the trouble on this occasion, frantic efforts by the engineers (always referred to in those days as mechanics) produced the required results within an hour. The increasingly impatient and fatigued pilots found this wait intolerable: nevertheless, lightly loaded for the short stretch of 787 miles to the finish at Flemington Racecourse, Melbourne, and twelve hours and nineteen minutes after first landing at Charleville, *Grosvenor House* was away for the second attempt at the last leg.

Now the pilots were showing serious signs of fatigue. They had intended to take half-hourly turns at flying and sleeping, but even these proved to be too long and before reaching the destination they were struggling with stretches of ten minutes apiece. The southern Australian coastline was somewhere ahead, but this relatively short overland leg seemed endless to the two weary pilots. Eventually, though, Melbourne appeared to starboard and Scott lowered the Comet's nose for a fast and dramatic finish, for which he was required to fly between two pylons at a height not above 200 feet. The red Comet G-ACSS *Grosvenor House* was first to cross the line and, unless someone along the route put forward any objections to the pilots' performance that could lead to disqualification, they knew that they were the winners.

For Scott and Black the race was over, but they were unable to land the Comet on the small racecourse and they flew on for the fifteen miles to the aerodrome at Laverton. Although for the victors the stresses and struggles of the flight were behind them, they had other pressures to face. For all the other contestants, though, still the battle was on. For the world at large, too, the unexpected rapidity of the flight from England to Australia — the total time taken had been only 70 hours, 54 minutes and 18 seconds — would be only the start of things to come. Certainly the prediction in *The Times* had been turned into truth.

With the race behind them, fortunately the winners had much to say and to put in writing. Already I have quoted some of Scott's remarks, but to balance the human side of the operation I end this chapter with opinions from his co-pilot. Black wrote two appreciations. The first, about the engine that kept them in the air: 'Seldom has any engine had such a gruelling test as our faithful starboard motor. We knew it, and our hearts seemed to beat in tune with its throbs.'

Then about his partner: 'It is difficult to explain the confidence one man may have in another, but without that confidence high speed, long distance flying by two pilots would be impossible. In the Singapore crisis with any other man than Scott at the controls, I should have died of heart failure'; and finally he wrote 'Scott has told you of our first and unalterable decision not to accompany any other pilot from Mildenhall to Melbourne. Surely events have shown we were right. That was the best bargain I ever made.'

Chapter 5
After the British Victory

When C.W.A. Scott and Tom Campbell Black landed at Laverton, the ordeal was far from finished. A. O. Edwards, the winning Comet's owner and financier, was there to greet them; and so were thousands of others, including Jean Batten of record-breaking flying fame. Very soon, however, still deaf from the hours and hours of droning of the aircraft's engines, numb and dazed, the joint victors were aboard a de Havilland Puss Moth to be flown back to Flemington for the start of a series of receptions and presentations. Scott, who said that the flight from England had been a lousy experience and 'that was praising it', was asked what his future ambitions were. He could think no further ahead than to express his need for a bath and a lot of sleep.

On arrival at Flemington, and without time for rest, Scott and Black boarded the first in a line of cars for a procession to the dais. There the Lord Mayor of Melbourne expressed Australia's appreciation that the race had been won by a British crew in a British aeroplane with British engines. Sir Macpherson Robertson, too, had much to say, mainly about bringing the peoples of the world into closer relationships and (optimistically perhaps) that this must lead to closer international friendships. These first orators said largely what would have been expected from them, but the third to speak created a pleasant surprise. He was Colonel Wright, the Netherlands Consul in Australia, who on behalf of KLM (whose competing DC-2 was still in the race on its final lap) presented the winners with a bunch of flowers and a card congratulating them on their navigation, ability and perseverance.

Now it was Scott's turn to reply. Understandably he was not in a mental or physical condition to win a prize for the best speech; he offered little more than expressions of appreciation for the unbelievably warm welcome that he and his partner were receiving. Although shortly after this the winning pilots were driven from the crowds for a few hours of relative relaxation, the reprieve was shortlived and soon reporters from the world's Press were all over them. In support of his earlier remark about the lousy trip, when asked which was the worst part of the flight he replied 'The whole lot'.

Messages of congratulation continued to arrive in all forms, including wires from King George V, the Duke of Gloucester, the Prime Minister (Ramsey MacDonald) and, expectedly, a telegram from Geoffrey de Havilland. As soon

Honour for the winning aeroplane: 'Grosvenor House' being towed through the streets of Sydney for all to see.

as they could get into print, the newspapers plugged the fact that everything contributing to the success was of British origin, down to the grapenuts cereal and the tomato juice that the pilots had consumed en route. The praise was not only from the British Press, for even the *New York Times* stressed the international significance of the British achievement.

While all the activities were under way in and around Melbourne, for others there was still work to do in the air. KLM's DC-2 continued to hold second place, but for the first time in the entire journey Parmentier and Moll were facing problems. They had arrived at Charleville 45 minutes earlier than had been expected and they were held orbiting until the ground party had established a flarepath. Understandably the Dutchmen were furious about this, but after an hour they were airborne again, to face the worst leg of the route. The weather was atrocious; cumulonimbus clouds generated electrical storms that were not safe to penetrate. Attempts were made to fly over the top, but these were unsuccessful, so Parmentier headed east with the idea of following the coastline round to Melbourne. Total darkness was interspersed with flashes of lightning that revealed ice building-up on the wings and the propellers. The crew mutually admitted that they did not know where they were.

Unlike the Comets and most of the other competing aircraft, the DC-2 had radio, with a wireless operator among the crew; an SOS message was received on the ground, but due to the electrical interference no reply was received on board the aeroplane. People rushed to help. Several small airfields were brought into action, with all available facilities laid on to form emergency services; on at least two sites, makeshift flarepaths were provided by car headlamps. Even the names of towns and railway stations were flashed continuously by signal lamps, while at some places the street lamps were turned on and off to signify the names in morse. Eventually the airliner arrived over Albury, which is about 190 miles north east of Melbourne; here the town's name was flashing and cars were lined up to made a makeshift flarepath on the racecourse. With the added help of the light from a flare released by the aircraft, Parmentier decided that he would land and he managed to do so safely on the storm-soaked surface.

During the remainder of the night the DC-2's wheels sank gradually into the mud. In the early morning groups of local residents dug and others tied ropes to the undercarriage; eventually muscle-power won and the large airliner was moved onto slightly less soggy ground. The take-off from a small, waterlogged racecourse was not expected practice for an airliner crew, but Parmentier decided that if all possible weight could be removed he stood a good chance of success. He knew that the only alternative would be to give up; clearly that was unacceptable. Until this stage three passengers and some mail had been aboard, but they were off-loaded and after a slow, ponderous run PH-AJU staggered into the air, to become the second aircraft to cross the Flemington finishing line after a total elapsed time of three days, eighteen hours and nineteen minutes.

Meanwhile Roscoe Turner in the Boeing 247D was having an uneasy time. Heavy rainstorms had forced him off track while skirting the eastern coast along the Bay of Bengal and he had made an unscheduled landing for fuel at Alor Star, an optional checking point on the north-west coast of Malaya. All was well for the relatively short stretch to Singapore, but about an hour after take-off from there the port engine began to lose oil pressure. He continued ahead on the other, depending on the continued reliability of the starboard Pratt and Witney Wasp, to land at Koepang on Timor Island. From here he needed to face the treacherous sea crossing, which created fear among aviators even without engine problems, but to tackle the trip with a good power unit on one side only called for a strength of determination that would have defeated many people. In normal circumstances, to attempt the flight would be grossly unwise, but this was a unique situation and not the time to throw away the chance of success.

The Boeing arrived safely at Darwin, but ran into navigational difficulties on the run to Charleville. Still cruising with the sick engine operating at greatly reduced power, Turner and Pangborn successfully received direction finding assistance from the Royal Australian Air Force and they were vectored to a point immediately overhead their destination, which sounds a remarkable achievement for 1934. On the ground at Charleville mechanics spent an hour in searching for the cause and then for a cure for the loss of oil pressure. The airliner took off again, with Melbourne as the target, but this time not only did the oil pressure fall again but a serious oil leak developed from the other engine, so NR 257Y landed at Bourke, in New South Wales, which lies almost due south of Charleville and about one-third of the way along the route. There was no chance of curing either problem, so in true style Turner opted to throw

ALL LETTERS SHOULD BE ADDRESSED TO THE COMPANY AND NOT TO INDIVIDUALS

THE
DE HAVILLAND
AIRCRAFT CO., LTD.

DIRECTORS
A. S. BUTLER CHAIRMAN
G. DE HAVILLAND
C. C. WALKER
F. T. HEARLE
T. P. MILLS
W. E. NIXON

STAG LANE EDGWARE
MIDDLESEX ENGLAND

TELEPHONE:
EDGWARE 0171 (8 LINES)
TELEGRAMS:
HAVILLAND
PHONE, EDGWARE
CABLES:
BENTLEY CODE

ASSOCIATED COMPANIES IN AUSTRALIA, CANADA, INDIA AND SOUTH AFRICA

GdeH/WS. 25th October 1934.

All of us must feel proud that a Comet has won the greatest race in the history of Aviation.

The wonderful success of Messrs. Scott and Black is due, not only to te untiring efforts of those who directly worked on the Comets, but also to the spirit of loyalty, enthusiasm and co-operation which every member of the Company displays. We sincerely thank you for your efforts in the past, and know that this spirit will carry us on to even greater successes in the future.

G. de Havilland.

CONTRACTORS TO THE AIR MINISTRY, DOMINION, COLONIAL AND FOREIGN GOVERNMENTS THROUGHOUT THE WORLD
PROPRIETORS OF THE DE HAVILLAND SCHOOL OF FLYING · HATFIELD · HERTS

ALL SALES ARE MADE SUBJECT ONLY TO THE COMPANY'S SPECIAL WARRANTY.

caution out of the cockpit and make a final dash to Flemington, hoping to beat his Dutch rivals; he was unaware, of course, that they were there already. Fortunately both Wasps stood the strain and the Boeing roared between the racecourse pylons to take third place across the line.

Now we return to Waller and Cathcart-Jones in G-ACSR. As they had experienced their share of problems on the earlier sections of the race they had no way in which they could make up for so much lost time. However, their ill fortunes did not persist throughout, for after their troubles at Baghdad and Allahabad and later their emergency landing at Batavia, with one exeption the subsequent runs appear to have been relatively free of problems. A servicing and fuelling turnround of only thirty minutes at Darwin sufficed to have 'CSR quickly on its way to Melbourne, but reported lubrication trouble (falling oil pressure?) caused the crew to make a precautionary landing at Mount Isa; this may have been a case of mistaken identity for Cloncurry, which was readily available as one of the optional check points. After this delay, though, 'CSR was away again to Charleville, to finish fourth, but the excitement of the achievement had died and only a handful of people were present to greet them. This must have been a discouraging experience after such a feat.

Although barely publicised, the aircraft to gain fifth place in the speed section of the race was of British origin and from the same stable as the Comets: an early production de Havilland Dragon Six (the original name for the Rapide) flown by New Zealanders J. D. Hewett and C. E. Kay, who managed to complete the route in twelve stages compared with the nineteen (eighteen planned) that had been taken by the much larger DC-2; the Boeing 247D had completed the course in eleven legs, ten of which were intentional. I mention this because the winning Comet and its crew had managed to cover the route of 11,300 miles in only seven stages, including the unintended landing at Kirkuk on the first leg from Mildenhall.

Many competitors had failed to finish and several had some dramatic experiences in their travels. This is not the place to recount the tales of all the pilots who were left behind, but I must mention the achievements of three crews in their light single-engined aircraft. In this field the most notable achievement must have been that by the Australian C. J. Melrose who in a de Havilland Puss Moth attained third place in the handicap section, flying average stage-lengths of more than 1000 miles in an aeroplane that cruised at no more than 105mph. Before the race he had achieved the fastest flight from Australia to England, in 8 days and 9 hours, beating times previously established by Rubin and Waller and by Mollison, but unfortunately Melrose's flight was not officially observed so no official record could be claimed. Also worthy of special credit were the New Zealanders M. MacGregor and H. Walker who, in their open-cockpit Miles Hawk Major, gained fifth place in the handicap race. Each of these light types depended wholly (and successfully) on the reliability and performance of the de Havilland Gipsy Major four-cylinder engine of only 120 brake horse-power. Fourth place in this section of the event went to British entrants D. W. and K. G. Stodart, in a British Airspeed Courier powered by a British Armstrong-

Siddeley Cheetah nine-cylinder radial engine. So the Comets and their engines were not the only British products to achieve success in the race, nor were they the only examples of de Havilland design and workmanship to do so.

The tumult was on the decline and both aircraft and crews needed to return to their own countries. Some planned to fly back, while understandably others chose ships to provide the best homeward paths. The two Comets were to follow very different travel patterns, for the pilots of G-ACSR planned a rapid turn-round in Australia to face an immediate dash back to England. A person with an imaginative mind must wonder how anyone could have the mental or physical stamina to face such an onslaught immediately after spending several days of strain in the cramped confines of the Comet cockpit, but Waller and Cathcart-Jones seemed to be in sufficiently fine spirit to tackle the task.

Reports say that after a brief rest in Australia, the crew of 'CSR took-off to return to England, but the picture falls into true human perspective only when we realise that they crossed the finishing line at Flemington on 25th October and departed for home on the 26th! Their aim, though, was not to stretch themselves or the aircraft to such fine limits of timing as had been necessary on the outward route; they would fly only by day and they would go to bed each night.

The flight home was not free from problems. During a ground-run prior to their intended take-off from Allahabad the earlier engine problem recurred and subsequent investigation revealed two cylinders in poor shape. The unfortunate Mollisons who had been stranded there on the outward attempt were still at Allahabad with G-ACSP, so sportingly Jim offered the necessary replacement parts from his still-sick machine.

After servicing and rest stops at Karachi and Baghdad, Waller and Cathcart-Jones intended to make only a brief pause at Tatoi Aerodrome, Athens, but here a pilot's biggest single bogey — the weather — caused the crew to give precedence to caution. Typically heavy Mediterranean storms were in action off the Italian coast and 'CSR's crew remained at Athens to await better conditions. Despite the delays, though, Owen Cathcart-Jones and Ken Waller arrived at Lympne aerodrome in Kent at 1.26 p.m on Wednesday 2nd November. Bernard Rubin, the Comet's owner who originally had intended to fly in the race to Australia but who had been unable to do so because of illness, was there to greet them. They had established an out-and-back record, with a time of 5 days, 6 hours and 43 minutes for the flight home and a total time away of 13 days, 6 hours and 43 minutes.

Most values must be relative and no-one would wish to criticise the Press or the public for the enormous credit bestowed on C.W.A. Scott and Tom Campbell Black for their achievements in winning the MacRobertson air race to Australia. However, such success tends to minimise the appreciation given to others and perhaps this is stressed most effectively in an editorial comment that appeared in the issue of *Flight* dated 8th November 1934. Under the heading '23,000 miles in a fortnight' the text includes:

'Generally speaking, this extraordinary effort on the part of Cathcart-Jones and Waller has not received due credit. After the excitement of the race itself, the world was inclined to forget them. Quite apart from the various records that they have broken, and the indication given of the possibilities of higher commercial speeds over the route, the two pilots have fought continuously against a series of misfortunes.'

This provides an appropriate opportunity to look at the results of the race in more purposeful depth than just considering who and what came first, second, third or fourth. Clearly great differences lay between the underlying philosophies of the de Havilland Aircraft Company and KLM Royal Dutch Airlines. Although each set sights on getting an aeroplane from England to Australia in the shortest practicable time, the two leading British entries were seen by most people to be based on sporting principles, using an aeroplane designed specifically for the purpose and *in itself* having relatively low potential in other spheres of aviation activity. By contrast, the DC-2 was an established and proven airliner already in successful revenue-earning service in the United States and the Dutch aim was wholly commercially based. Indeed Parmentier and Moll refused to consider themselves as racing pilots and they accepted their task wholly as a long-range route-proving trial; but to achieve their aim they could not ignore the active competitive element and clearly their minds were occupied with thoughts of Scott and Black in the same way that the KLM airliner's progress was much more than just a passing spur to the Comet crew.

Both approaches proved their success. As a direct result of the KLM achievement, the airline ordered ten more DC-2s and immediately planned a drastic expansion of the route network. This move was not entirely popular with many people in the Netherlands, for the Dutch manufacturing industry was in business building a range of Fokker airliners, but even the best on immediate offer from the home team was unable to match the performance of the American machine.

Whilst we look upon the Comet as an aeroplane designed to win a race, it should not be thought of technically as a racer. In the past few years the misnomer has been carried to the extreme and the term Comet Racer (the capital R implying that this is the aircraft's type name) has been perpetuated. Clearly this inaccuracy has been brought about with the world's best will, presumably to differentiate between the DH 88 Comet of 1934 and the later DH 106 Comet airliner, but for the sake of historical accuracy perhaps we can give this mistake its final burial. Although the use of this title in post-war years may be understandable, surprisingly the error is not new, for before the end of the year in which the first three DH 88s were built, *Flight* saw a need to clarify the position.

Under the sub-title 'Not a Racer' three paragraphs read:

'The Comet is generally referred to as a racing aeroplane, and to the extent that it was designed for the Australian Race, and has not a large and commodious cabin, this is true. It differs, however, very greatly in other respects from a racing machine. The operational conditions (which it is believed were adhered to) were that full throttle at 10,000 feet or over should be employed. This means (at the r.p.m. allowed by the propeller)

AIRCRAFT D.H. 88 "COMET" No. 1996, G-ACSS

DATE. (1)	HOUR. (2)	TIME IN AIR. (3) HRS.	MINS.	TIME SINCE OVERHAUL. (4) HRS.	MINS.	REPAIRS
Brought	Forward					
9-10-34			20			Test Flight at Hatfield. Capt Broad.
12-10-34			25		45	do. do
12-10-34			20	1	05	do do
12-10-34			15	1	20	do Mr C. Scott
12-10-34			15	1	35	do Mr T Campbell Black.
12-10-34			15	1	50	do Capt Broad
14-10-34			20	2	10	-do- Mr C Scott & Mr Black
15-10-34			40	2	50	do -do-
16-10-34			10	3	0	do Mr Black
17-10-34			30	3	30	do Mr Scott & Mr Black.
20-10-34		70	59	74	29	England to Australia Mr C. Scott & Mr T. Black.
Carried	Forward	74	29	74	29	

The first page from the logbook of G-ACSS, showing the race from England to Australia recorded as one entry.

that 75 per cent or less of the type-test power for continual operation was used in the Race. The machine also complied with the International Airworthiness conditions on its full load.

A further point is that in using only three-quarters of the "continual operation" power, a specific consumption could be used which was much lower than has ever been contemplated for racing purposes.

Perhaps the forced landings by night and the full-load landings by night which took place during the race might also be adduced to show that the rules of the Race do tend to exclude what is generally understood by the term "racing machine." '

Possibly this could have been explained more clearly, but what was clear enough is the aim of the message: that the DH 88 Comet was not just a racer in the usual meaning of the word. That the winning Comet subsequently served for a year in the Royal Air Force (described in the following chapter) adds more weight to the argument that in technical, if not in practical, terms, the aeroplane had rather more to offer than would be expected of a specialist racer.

Elaborating a little on the operating height of 10,000 feet, we may be able to pin-point some of the problems with the Gipsy Six R engines. Incorrect mixture settings have been stated as the most likely cause of the troubles, but the pilots could not be blamed as they were unable to gain practical experience in the short time between their first flights and the start of the race only a week or so later. After levelling-out at the top of the climb, the procedure followed throughout was to weaken the mixture until the rpm began to reduce and then ease the lever to a very slightly richer position in order to regain the original power output. For obvious reasons of economy, the tendency was to err on the weak rather than the rich side and the engines seemed to suffer from the symptoms resulting from long running in weak mixture. One report of the period anticipates the eventual introduction of 'automatic devices' that would relieve pilots of the work of mixture setting. Overheating was another engine problem, as in the quest for speed the cooling intakes had been made too small to be effective in tropical temperatures. The intakes were enlarged later in the Comet's career.

When we view the Comet as an aeroplane in terms of its technical potential, the gap between the first two aeroplanes to reach Melbourne begins to reduce in size. Geoffrey de Havilland may have been a sportsman with a strong desire to see a British product hit all the headlines by winning the race to Australia, but equally he was a businessman. The Comet cost his company a lot of money and this direct loss to the firm's bank balance was not accepted on philanthropic grounds.

The de Havilland task was to design, build and *sell* a very broad range of aeroplanes to customers all over the world: the DH 82 Tiger Moth military and civil trainer; the DH 83 Fox Moth with its enclosed cabin for four and its pilot alone behind with his head in the breeze; the DH 84 Dragon mini-airliner with accommodation for six passengers (the type that had been used in 1933 by the Mollisons for an attempt on the world's long-distance record); the DH 85 Leopard Moth three-seater for the private owner; the DH 86 four-engined biplane airliner; and the DH 89 Dragon Six, later to become the most famous short-haul feeder-liner — the Rapide. All these designs of DH origin were in various stages of production or development and the prestige and publicity from success with the Comet could do more good to the sales cause than any number of routine advertisements.

So for the longer term the people behind the Comet and those behind the DC-2 had broadly similar aims in mind — to improve their respective businesses. All the Press publicity that surrounded the successful pilots and their aircraft and all the sounds of exuberant enthusiasm that came from the lay public rubbed-off very noticeably onto two marked examples of strongly-guided private enterprise. The name and the achievements of Geoffrey de Havilland were known to all, but many readers now, as then, may wish to be reminded that despite its title of Royal Dutch Airlines, KLM was a private company, operating under the strictly frugal eye of its founder and chairman, Albert Plesman. He spent money only when he could foresee a healthy financial return, and although in many ways he was as progressive as he was aggressive in his endeavours, for several years after the availability of radio stations on certain routes, he had refused to equip his fleet with such a novel device as wireless, as it showed no direct sign of increasing his revenue.

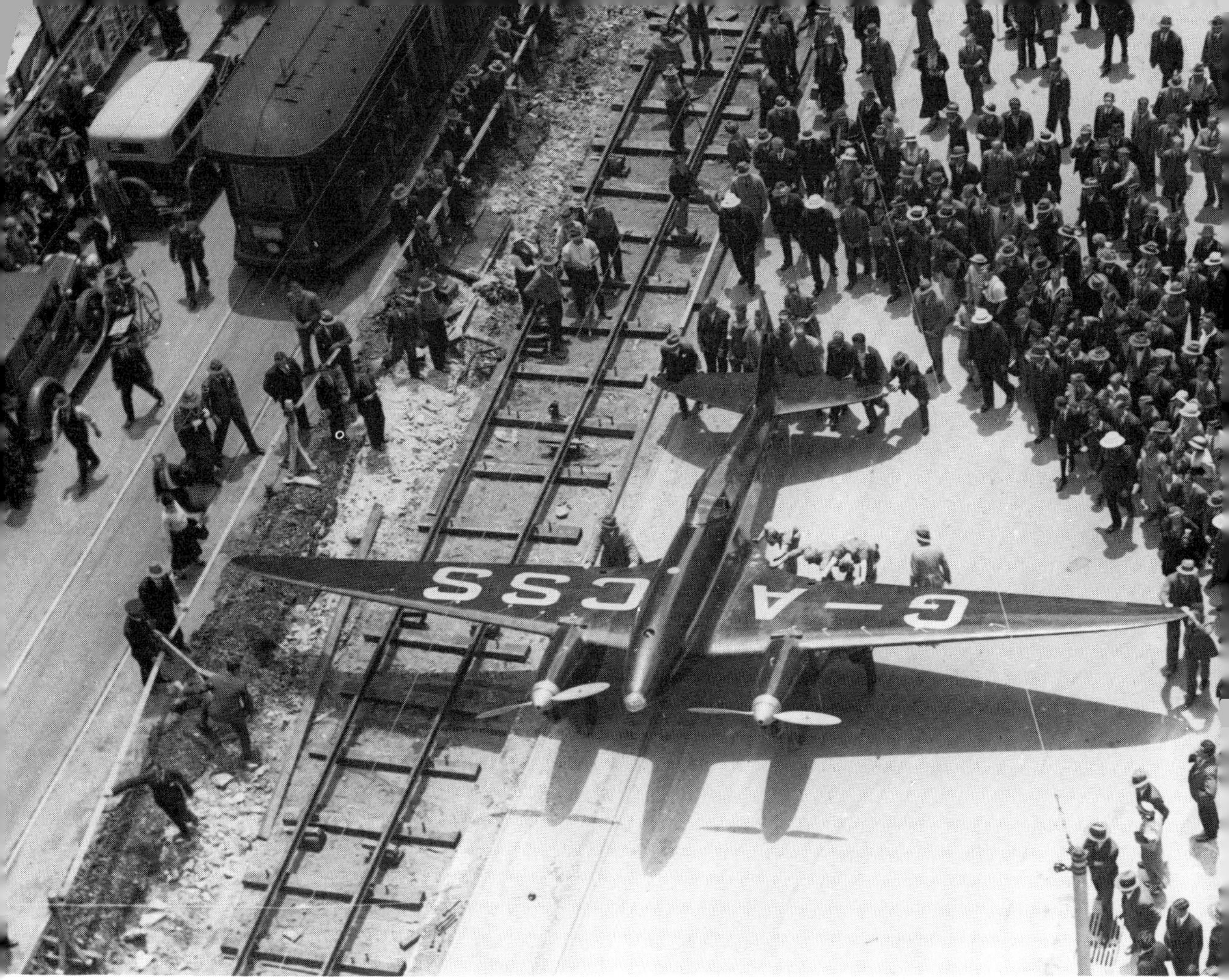

The winning Comet prior to loading at the docks for the journey home to England. 'Grosvenor House' travelled, fully assembled, as deck cargo.

Nearer home, various people involved either in running airlines or building aircraft for them were tending to lick their wounds and make defensive statements. British and other European airliners were several years behind those operating in the United States, and while clearly the Comet's performance in itself could not be levelled against the poor scheduled timings of the regular passenger and mail operators, the fact that any aeroplane had reached Australia from England in fewer than three days must have been an aggravating fact to accept. Even if the Comet was discounted as just a fast aeroplane with no load-carrying capacity, certainly the performance by the American-built and Dutch-operated DC-2 could not be good for the digestion.

In this context the General Manager of Imperial Airways stated that, as a transport contractor, his organisation was prepared to provide whatever

'Grosvenor House' arrives home.

services, at whatever speeds and capacities, were required by its clients. From the Managing Director of Short Brothers Ltd., came the statement 'If commercial airline operators call for aircraft in which the primary objective is a very high cruising speed and other desires are made of secondary importance, the demand can be met by British aircraft constructors without any difficulty. There has been no such demand from aircraft operators in this country, at least, from those companies who have had the most experience in operating airlines.' Frederick (later Sir Frederick) Handley Page said 'Because a race has been run and an extraordinary demonstration given of engineering skill and pilots' stamina in navigation, it does not mean that similar speeds should be aimed for as essential in commercial aviation.' True to business form, and much to his credit, he added that low landing speeds were far more important as essential adjuncts to safety; five years previously he had produced the Gugnunc, which with its full-length slats was capable of controlled flight at only 33.5 mph and had been runner-up in the Guggenheim air safety competition in the USA!

Shortly after this, *Flight* joined in the mêlée of words. In the commercial aviation column for 1st November 1934, "A. Viator" said: 'People are asking if speed is so terribly expensive as some experts would have us believe, and if the fact of doing twice as much work in an hour is not as important in the air as in a factory, for example. The truth probably lies in the fact that speed will pay on long distances but not on short routes.' Then the same journal went on to weigh heavily against commercial night flying.

By 8th November the Chairman of Imperial Airways joined in the 'club' of British experts who saw no future in seeking to fly faster, yet within a few sentences he announced with obvious pride that his organisation was planning a regular service from Croydon to Sydney that would take only 7¼ days. It could not be self-financing and the British Government had agreed to a subsidy . . .

During this peak of discussion about speed, many people had stressed the advantages to be gained by operating fast air-mail services throughout the world; and by December of 1934 Imperial Airways started the first all air-mail route from England to Australia. That cannot be a coincidence.

Was it the Comet, or the DC-2, or both, that spurred the airline industry into action? Almost certainly if there had been no MacRobertson air race and if such high-speed timings had not been achieved along the route, several more years would have drifted past before anyone had introduced these marked accelerations in the regular services.

Chapter 6
More Achievements – and more Comets

Without Sir Macpherson Robertson and the race from Mildenhall to Melbourne, there would have been no DH 88 Comet and the successes of G-ACSS and G-ACSR in the great event must hold first places in most people's minds. However, Comets and their crews went on to attempt and often to achieve many more records and, on the strength of numerous high-speed flights over long distances, the aeroplane was studied seriously as a contender for other specialised but more workaday uses.

Because the record return run from Australia to England by Ken Waller and Owen Cathcart-Jones in 'CSR was inseparable from the outward flight that gained them fourth place in the race, this notable achievement has been described already in Chapter 5. Although in itself the journey home was significant, the most important reason for this flight's fame was the short time taken to fly from England to Australia and back again with barely a break at the far end.

While 'CSR was gaining a new record, 'CSS and its pilots were ready to rest for a while on their laurels. There was no race homeward. The Comet was displayed at several sites in Australia and no one on that side of the world wished for it to leave. Eventually, however, *Grosvenor House* returned on a leisurely ship-borne voyage and docked at Hull. Among the people who have contacted me with helpful information is Stan Baker, one of two de Havilland engineers consigned to the task of removing 'CSS from the vessel, on which it had travelled free of charge as deck cargo; unfortunately the Australian dockers had just slung ropes round and these had caused damage to fairings. The aircraft was dismantled on the dock finger in freezing conditions and then was taken to Yeadon aerodrome (now Leeds/Bradford airport) for repairs and to have special gear fitted for road transportation.

For a while *Grosvenor House* was in the limelight again, for a national newspaper sponsored its display at several of Lewis's stores in northern towns; Mr Baker has vivid memories of the anxious moments when moving the wing and fuselage on slings through high-level windows, (one of which had been especially enlarged), with his colleague Mr Scott-King standing on the wing centre-section to direct operations, for which each was awarded a very special bonus of £15. This was in Liverpool, where a luncheon was laid-on to honour

After docking at Hull, 'Grosvenor House' was dismantled on the dock finger and transported for display in several of Lewis's stores in the north. Here it is at Liverpool, where, because of construction work hampering parking, G-ACSS was lifted ***over*** *the building. An engineer stood on the wing during this precarious operation!*

the winning pilots. In fact, both were in demand as speakers. Usually Scott spoke first and gave dramatic accounts of their experiences, while Black added some quiet humour to each occasion. At least once he said "It is right that Scott should speak before me. He crossed the finishing line first. He was sitting two feet in front of me!"

Although the name and the reputation remained in the minds of many people, for a time the winning aeroplane seemed unloved. Tom Campbell Black was planning some more record flights and he asked A. O. Edwards if he would sell him *Grosvenor House,* but he refused. I learnt this from Mrs. C. F. Hughesdon (at that time the noted actress and impersonator Florence Desmond, who married Black in 1935) and she provided me with many other items of information; apparently neither Scott nor Black received any of the prize money, nor did they have the gold cup. Apart from a generous cheque from Lord Wakefield, whose oil had been used throughout the race, the only financial rewards that either managed to glean from their achievements were payments for newspaper articles and for allowing their names to be used in commercial advertising. After Edwards had refused to sell his Comet to Black, he disposed of it to the Air Ministry for £7000.

Almost immediately after the Australia race and well before *Grosvenor House* had returned home, 'CSR was back in the high-speed business. Renamed *Reine Astrid,* on 20th December 1934 the green Comet left Brussels for the Belgian Congo carrying 300lbs of Christmas mail. Ken Waller was on board again, but this time the second seat was occupied by Captain Maurice Franchhomme; they arrived at their planned destination after 22 hours 35 minutes in the air, but not satisfied with this they returned from Leopoldville with 4000 letters on board, to achieve the out-and-back journey of 8,882 miles in flying times totalling only 44 hours and 29 minutes.

Very shortly after returning to England from Australia, G-ACSR was named 'Reine Astrid' and on 20th December 1934 left Brussels with Christmas mail for Leopoldville in the Belgian Congo. The out-and-back flight of 8,882 miles was completed in a flying time of 44 hours and 29 minutes.

Almost immediately, 'CSR was bought by the French government and registered F-ANPY. It was overhauled by its makers at Hatfield and repainted in a red and white colour scheme; initially, however, the French refused to accept it, for they insisted that it was not capable of achieving the promised range. This was because of allegedly high fuel consumption on a series of short check flights, but the problem was traced to fuel sloshing about and syphoning on the climb due to the absence of baffle plates in the tanks. A modification to the venting system put that right and on the delivery flight to its new owners F-ANPY beat the London to Paris record, landing at Le Bourget only 56 minutes after leaving Croydon. On this ferry trip the pilot was Hugh Buckingham; with him was Martin Sharp, who later became well-known as the de Havilland historian and who in 1960 wrote the well-respected book *'DH'*.

TIMES AND SPEEDS OF BRUSSELS-LEOPOLDVILLE-BRUSSELS FLIGHT BY G-ACSR

OUTWARD JOURNEY			*Time of flight hrs. mins.*	*Distance*	*Average Speed*
20/12/34	Brussels Oran	10.48 15.44	4 56	1118 miles 1798 km	227 mph 365 kmph
21/12/34	Oran Niemey	07.30 15.30	8 00	1673 miles 2695 km	209 mph 336 kmph
22/12/34	Niemey Point Noire	03.20 11.30	8 10	1525 miles 2455 km	187 mph 301 kmph
	Point Noire Leopoldville	13.10 14.50	1 40	250 miles 402 km	150 mph 241 mph
			22hrs 46m	4566 miles 7350 km	200.5 mph 322 kmph
HOMEWARD JOURNEY					
26/12/34	Leopoldville Niemey	04.00 11.40	7 40	1525 miles 2455 km	199 mph 320 kmph
27/12/34	Niemey Colomb Bechar	06.40 14.25	7 45	1373 miles 2210 km	177 mph 285 kmph
	Colomb Bechar Oran	15.30 17.05	1 35	300 miles 482 km	190 mph 306 kmph
28/12/34	Oran Brussels	07.20 12.14	4 54	1118 miles 1798 km	228 mph 367 kmph
			21hrs 54 m.	4316 miles 6945 km	197 mph 317 km
TOTAL TRIP			44 hrs 40 m.	8882 miles 14295 km	199 mph 320 kmph

N.B. These are figures from the DH archive. The times differ slightly from those recorded by the Royal Aero Club.

Early in 1935 the former 'Black Magic' lost its British identity as G-ACSP, to become CS-AAJ 'Salazar' of the Portuguese Government.

The third of the Comets that participated in the MacRobertson race — G-ACSP *Black Magic,* owned and flown by the Mollisons — was offered for sale as soon as it returned to England after its technical ordeal at Allahabad. It was purchased by the Portuguese Government, registered CS-AAJ, named *Salazar* and in February 1935 was flown the 1010 miles from Hatfield to Lisbon in 6 hours 5 minutes. The pilots were Carlos Bleck and Lieutenant Costa Macedo. Two years later *Salazar* was damaged and returned to its makers for repair. Afterwards, on 2nd July 1937, Macedo, CFI of the Portuguese National Flying School, was aboard again and the flight time back to Lisbon was reduced to 5 hours and 27 minutes. Throughout its Portuguese service 'AAJ retained its all-black finish.

A strange feature of DH 88 history is that by late spring 1935 there was no Comet on the British civil register. With 'CSP in Portuguese ownership and 'CSR in the hands of the French, the time came for 'CSS *Grosvenor House* to lose its original identity. Unlike the first two, though, it remained on home territory, exchanging its sporting red-and-white colour scheme for all-over silver, with RAF roundels and the Service serial K 5084. After returning by sea from Australia, (by which time the logbook showed a total of 70 hours 59 minutes flying) it was not flown again until 21st June, when it was test-flown in its new military markings by Hugh Buckingham and delivered to the Aircraft and Armament Experimental Establishment at Martlesham Heath. Only five days later it was flown to Hendon, to participate in the famous air pageant on 29th June. After the show the military Comet returned to Hatfield for six weeks, but the next airing was a test-flight on 10th August, *after* which the logbook shows some more work carried out including repairs to the port wing and a complete repaint! Eleven days later it was test-flown again and survived only two more flights from Martlesham before the next trouble arose.

Away from the civil scene and despite flying fewer than 33 hours during its eighteen months in Service ownership, K 5084 endured a life of considerable turmoil. On 30th August the troubles with undercarriage retraction and

Shortly after its return to England, G-ACSS temporarily lost its civil identity to become K 5084 in the Royal Air Force.

lowering caused the machine to be damaged by landing with its wheels only partially down. The engines were removed, the aircraft was dismantled as far as is possible with a one-piece wing and by 3rd September it was back at Hatfield yet again. Three local test flights in December and January, a delivery to Martlesham and more trouble; on 7th February 1936 K 5084 was returned

The sole military DH 88 – K 5084 – participating in the RAF's Hendon Air Pageant.

by air to Hatfield, where the front petrol tank was replaced and the other two were repaired.

There is no need to study the records very deeply to gain the impression that the Service and this 'hot' aeroplane were not the best of partners. Although cleared as serviceable after a flight test on 26th June, the Comet was not collected from Hatfield for nearly a month, this time going to the Royal Aircraft Establishment at South Farnborough. Here several known names such as Flight Lieutenants Atcherley and Purvis appear on the test forms for a series of noise measurement trials, while on 13th August Flying Officer Clouston appears on the Comet's records for the first time. Later in this chapter and again in the next, Arthur Clouston features frequently as the person who wrote more about the Comet and its flying characteristics than any other pilot. In fact, the only other detailed handling report that I have seen was written by a Frenchman, based on many working flights in F-ANPY and F-ANPZ.

Not all records are right and when two different reasons are given for an aeroplane making an unscheduled landing in a field, something, somewhere must be adrift. A logbook entry for K 5084 on 19th August 1936 reads:

> 'Delivery from Farnborough to Martlesham: 1 hour. Pilot lost. Precautionary landing at Butley. Flown out same day by Flt. Lt. Purvis to Martlesham: 10 minutes'

The book *Men with Wings* contains a section about Bruin Purvis, who later was awarded the D.F.C., A.F.C. and bar. The story says clearly that on this delivery flight both engines stopped due to fuel starvation. The resulting forced landing was successful. The pilot then studied the fuel system, paced and examined the field and declared to himself that all was in order for a take-off; that is, except for the fact that one propeller refused to be pumped into fine pitch. The tale describes the first attempt as a trial run, the next as a development of it and that neither was a very convincing demonstration. However, the third time was lucky and the Comet staggered away under asymmetric power to reach Martlesham late but still in good order.

The next fortnight saw the only time in the Comet's existence as a military machine in which it flew several times and completed 12 hours in the air on local test flights from Martlesham with no technical or other troubles other than wearing out the shoes of the tail skid. That is, until 2nd September, when K 5084 ended its time in the Service with a severe landing accident in which both undercarriage legs collapsed and the main fuel tank fell through the bottom of the fuselage. The civilian logbook of G-ACSS (which was maintained during the period of Service ownership) states 'believed overweight landing' while the writer of the official military report on K 5084 takes the trouble to put in the first paragraph 'the total weight was 5,000 lbs, this being 550 lbs less than the full authorised weight for the aeroplane.'

As with so many other instances of history, we are unlikely to learn more now, but whatever the cause, the reason or the excuse, this unpresentable arrival was a blessing. On 2nd December 1936, K 5084 was written off the Service records and when repaired G-ACSS would be able to emerge again

This could have been the end. K 5084 – alias G-ACSS – after its final landing in the hands of the Royal Air Force.

with its true and original identity. The aircraft's total logged flying time since new was 103 hours and 25 minutes.

While the most famous Comet of all was performing in its temporary role with the Royal Air Force, two more DH 88s were built. The first of these held no British registration, for it was ordered and bought by the French government as a partner to F-ANPY. Registered F-ANPZ, the new Comet had a mail hold in the nose in place of the landing light and, instead of the Ratier propellers, de Havilland/Hamilton variable-pitch units were fitted. These were modifications of the propellers intended originally for the first three Comets, but which could not be suitably developed in time for the MacRobertson race; they were the start of the range of long-serving D.H. propellers that performed so well on many aircraft in later years, but initially they were of American Hamilton design built under licence in Britain. On 5th July 1935 Hubert Broad, then de Havilland's chief test pilot, delivered F-ANPZ from Croydon to Le Bourget and in doing so knocked six minutes off the London-Paris record.

In April 1935, G-ACSR left its home country to become F-ANPY, as the first of two Comets that flew successfully for more than three years with the French Government. Here F-ANPY is taking off from Croydon on the delivery flight, on which it broke the record from London to Paris.

The cockpit of F-ANPY, with additional instrumentation squeezed into the cramped space. The large wheel on the right-hand cockpit wall is for retracting and lowering the undercarriage.

The French Government's purpose in buying two Comets was to evaluate them for possible use on a high-speed mail route over the South Atlantic. Already French-built Caudron C.640s had been ordered for fast mail services and some people considered the Comets to be mainly a spur to the home industry to speed-up development and production of the indigenous machines. Both Comets were based at Villacoublay, where all civil and military aircraft were tested. Surprisingly, they proved to be very reliable and, apart from a damaged wing once, following a tip-stall on landing, there were virtually no troubles during more than three years of continuous service. They were used for pilot conversion training and occasionally for flying VIPs, but the Comet's record-breaking potential was not ignored: on 1st-2nd August 1935 Jean Mermoz flew F-ANPY from Paris to Casablanca and on to Dakar in 15 hours 41 minutes at an average of 193 mph for the 2,990 miles, achieving an even better result on the return journey at 208.5 mph. Shortly afterwards Mermoz averaged 201.3 mph on a flight from Paris to Algiers and back in 8 hours 38 minutes. In 1938 both French Comets were flown to Istres (now the flight test centre), and after the Germans arrived in 1940 both were destroyed in a hangar fire. They had not been used for the planned mail flights, but they appear to have worked successfully in other roles and possibly they completed more flying time than any of the other Comets.

The fifth and final DH 88 was built primarily because A. O. Edwards had refused to sell G-ACSS to Tom Campbell Black. Immediately after the Australia race Black had started to plan several more long distance flights with the aim of establishing out-and-back records. He was seeking a new backer, so Florence Desmond, whom he was about to marry, approached Henry Horne, who threw regular parties that were attended by many rich people. Within a few days Mr. Horne came back with the news that a Sheffield stockbroker, Cyril Nicholson, was very interested in the possibilities and wanted to meet the pilot who planned to achieve so much. Black explained that he intended to fly to Capetown and back, to New York and back and to Hong Kong and back, each return journey to be completed in little more than a long weekend. The financier was impressed; he agreed to support the project, gave Black a cheque for the deposit on the Comet and construction began.

While the Comet was being built, plans for the flights developed. Black had chosen Gordon McArthur as his co-pilot, not because of his flying experience, but because the new aeroplane was to have radio and McArthur was a skilled wireless operator. Most of the details of the projected flights were worked out at Florence Desmond's house in the country near Bishop's Stortford. Here the Comet's colour scheme was decided as sky blue and silver and, in the hope that it would always come back, the new aeroplane's name was agreed: *Boomerang.*

One afternoon a heavy thud came from one of the flower beds and Gordon MacArthur was seen picking himself up and brushing the mud from his trousers. He exclaimed: "It's all right. I'm not going crazy. I've just jumped out of the bedroom window." He had read that the last part of the fall of a parachute jump is rather like jumping from a window ten feet high and as the

The fifth and final DH 88 to be built, G-ADEF, emerged in July 1935. Here on 8th August it is being named 'Boomerang'. Facing each other are Tom Campbell Black and Mrs. Black, formerly Florence Desmond, the well-known actress and impersonator.

two crew members had decided to be equipped with parachutes on the record-seeking flights he wished to know how it would feel if he should be compelled to use one.

By July the aeroplane was completed. On 8th August, registered G-ADEF, *Boomerang* was launched at Hatfield by Lady Fielding from Sheffield. Black and McArthur wasted no time and were away at 5.30 p.m. on the same day. The heavily laden twin swung badly on take-off, but soon was airborne and heading for Cairo, with the aim of a stop at Kisumu and then on to the Cape, which was 7300 miles away. The Comet reached its first destination in a record time of 11 hours and 18 minutes, but all was far from well. One engine had suffered seriously en route and on arrival was found to be almost burnt out. This was due to a shortage of oil, because before leaving England the level had been checked with a dip-stick for a DH 84 Dragon, which indicated the Comet's tanks to be topped when they were far from full. Engine spares and mechanics to instal them were flown rapidly from Hatfield and on 11th August 'DEF was ready to fly from Almaza back to England. By Black's standards this was a relatively leisurely flight in time terms, for neither the aircraft nor the engines behaved as well as he would have liked, but still the result was a record for Cairo to England of just over thirteen hours.

When the aeroplane had been thoroughly overhauled, on 21st September 1935 Black and McArthur started again. They were worried, firstly because they had missed the time of year that they had chosen originally for tackling the Cape record and secondly because delay might give the opportunity for others to come forward with aircraft that could tackle the route more rapidly. According to Florence Desmond's records, neither crew member really intended to wear a parachute, but by doing so the insurers agreed to reduce the premium and, for more humane reasons, she persuaded them to 'play safe'.

On this second attempt, *Boomerang* battled its way through and around several heavy thunderstorms, but arrived safely at Cairo with no technical troubles. The next stage was the leg for refuelling at Kisumu and this is where the real problems arose. For far too long there were no reports about the Comet's whereabouts and newspaper headlines declared 'Campbell Black missing.' By this time Miss Desmond was Mrs Black and I am very grateful to her for enabling me to find the full facts so that the true tale can be told.

Most records state merely that the crew of *Boomerang* experienced propeller trouble on the route from Cairo to Kisumu and, as a result, they baled out. I might have accepted this as the only information available, but fortunately at the time the *News of the World* published an article by Tom Campbell Black and to put right the record I feel that readers should see the full story:

> 'I am cabling from Atbara. You should see Atbara, and live in it for a day. A day would be enough. They tell me it is one of the hottest places on earth. Undoubtedly they speak the truth.
>
> But Atbara is paradise. At least, it was paradise to McArthur and me when we came rolling and jolting in on a camel-train last night. The camels are going to Kassala for dispatch to Eritrea. The Italians want them in their preparations for war.

'Boomerang' on the move at Hatfield, prior to the attempt on the record to the Cape, during which the crew were forced to bale out in most dramatic circumstances.

Well, we've had another go to get to Capetown and back quicker than anybody has done the double journey yet. Again we've failed. Distressed, of course. Who wouldn't be?

Finished with records? Not on your life. As a matter of fact, Mac and I are coming home hoping to find backing for another attempt.

Optimism? Effrontery? Call it what you like, but we are hoping just the same.

When we got to Cairo in eleven hours and ten minutes, and the beautiful machine, now an ugly, twisted pile in the African scrub, was answering perfectly to every touch and turn, I was confident of hitting Capetown in well under two days. So much for cocksuredness.

The only satisfaction is the knowledge that I have just stepped out of the most adventurous experience of my career.

On the very night of our departure we struck enormous and impenetrable thunder-clouds, and electric charges flashed across the *Boomerang* from one metal propeller to the other. They lit us up as though we were on fire. Solid sheets of water swirled around and hemmed us in, so that we could see nothing for hours, but we groped our way successfully over the Mediterranean, and went down into Cairo ahead of time.

Great bare patches along the wing of the monoplane, where the force of water had cut under the paint-work, were eloquent of the conditions through which we had flown. Rapid inspections of the motors at Cairo and an even quicker refuel enabled us to leave a few minutes in front of schedule, and we headed south on the second hop to Kisumu. Our course would have taken us right off the beaten track, but after two hours and a half the Boomerang's port engine began to lose revolutions. McArthur immediately got into wireless touch with an Imperial Airways' machine flying down the Nile, reported the trouble, and asked at which Imperial Station we were most likely to obtain quick repairs.

Possibly the responsibility of advising a suitable aerodrome for landing the heavily-laden Comet required considerable thought, and, as time slipped by without an answer, I got McArthur to wind in the aerial and help me at the controls. The machine at that stage required a great deal of force to maintain height.

We decided then to turn towards the river and make for Khartoum. Suddenly the starboard motor developed trouble also, and we could no longer keep our altitude. Gradually the plane dropped to 4000 feet. We hoped, however, to remain airborne long enough to make the journey, but from here onward we were on constant watch for more open country, where, if imperative, we might pull off a landing.

While passing over heavy bush the motor suddenly lost several hundred more revolutions, attributable, I suppose, to the strain imposed by flying in a semi-stalled condition with insufficient revolutions.

Then something went on the starboard propeller, and the vibration momentarily suggested that *Boomerang* had broken in two.

In the hot, dry atmosphere she continued to drop at over 110 mph and still nothing below offered the faintest possibility of landing. Bush was everywhere.

I write the simple truth in saying that we nearly stayed too long before deciding on our parachutes. Remember, we were rapidly getting lower all the time. Finally we prepared to release the cockpit-cover and get out.

Our decisions had already been made. I was to stall the machine, so that McArthur could jump clear, and avoid being fouled by the rudder or elevator behind, and I was to follow him as soon as practicable.

Here, however, our plans went amiss. Immediately the cockpit-cover was released it thundered back into the tail-mounting, and the terrific

The two Comets for the French Government, with F-ANPZ under construction. Any photographs of this machine in its final finished form are very elusive.

G-ADEF with the starboard Ratier propeller in fine pitch, indicated by the disc protruding ahead of the spinner.

inrush of wind, hitting both of us full in the chests, forced us down in our seats. At the same time, the machine, which I had stalled in order to get clear, whipped into the most vicious spinning nose-dive I have ever experienced. In that fleeting second, as *Boomerang* hurtled to the desert, we were both of us trapped and held by the inrushing air, and, strive as we did, we could not escape.

I caught sight of McArthur struggling desperately to clear the starboard side, but the wind drove him down, and finally threw him head-first out of the other side. I saw the ground rushing towards me then, and made a last, desperate effort to get out. Hazily I remember noticing that one shoe, momentarily caught in a control, had been ripped from my foot, and then in the next second I felt myself lifted backward by the wind. To my astonishment I was out.

McArthur, I understand, made his forced descent in full accordance with the instructions given by the makers. He tells me that he waited a while in order to get well clear before pulling his ripcord, and so was able to view from above the final explosion as the Comet struck the desert. It was shattered to pieces, and some of the pieces landed a hundred yards away.

My own parachute operation was far less in keeping with the book. I had been trapped, as I have said, until desperately near the earth, so that almost before I left the cockpit I had pulled the ripcord.

I was on my back head-downward when I felt the most astonishing tug at my shoulders. As I turned right-side up the machine hit the ground.

Almost immediately I too 'touched' with surprising gentleness, and the parachute-cords had no time to set me swinging like a pendulum. With McArthur, however, it was different. As he came down he was slung against a tree and knocked unconscious.

Immediately after freeing myself from the harness I had a horrible impression that McArthur was in the wreckage, because the thought flashed through my mind that as he got out first he should be down before me.

Actually, of course, by getting out first he was above me all the while, but in that silent moment after the crash I did not think of looking skyward.

My landing-spot was very near the smashed machine, and as I rushed towards it I saw Mac practically dead overhead, slowly coming down. While he lay unconscious I am afraid I was so relieved that both of us were safely out that I did not worry at all about his little argument with the tree, but waited quietly for him to recover. Anyway, his first words reassured me completely.

He expressed great relief at seeing me, and regretted that our bottle of mouthwash had gone. He had bitten a lump out of his tongue in the impact.

Before we could decide our next move a very surprised native lumbered up on a camel, followed shortly afterwards by others. Then began a trek to their village.

Here we rested until other camels were forthcoming, and we were taken a step nearer the Nile.

Mac and I spent the night in the open, and the following morning reached the water and obtained a boat.

The Nile is now in flood, and we were pushed and punted through marshy swamps until finally taken out into open water.

Another tiring ride, and we came to the railway-track at Cabushiya. Here we telegraphed, and ultimately caught the camel-train for Atbara.

The people here have asked constantly about our future plans. These, of course, are uncertain at the moment, but, as I said in the beginning, we hope to find someone high-spirited enough to back us for a further attempt to lower the record to the Cape.

It can be done, and by putting these high-performance machines through such gruelling tests a great and rapid development in aircraft can be achieved. Until I see you in England, au revoir'.

A Ratier two-pitch propeller being set into fine pitch with the use of a bicycle pump.

Through the freak weather conditions and not because of the fault of the aeroplane, *Boomerang* had not lived up to its name, but the crew were safe. The sponsor had not insured the machine, so instead of gaining any kudos for his aeroplane achieving a series of records, he lost the £10,000 that it had cost him. Very sportingly, though, he is known to have said "Tom and Gordon are alive and unharmed. That is all I care about." So, following an active life of only two months and probably a total time in the air of only about sixty hours, the last DH 88 to be built was the first to come to grief.

With F-ANPY, F-ANPZ and CS-AAJ working in continental ownership and G-ADEF a write-off, the future for the DH 88 as a British registered aeroplane fell entirely to G-ACSS, which alone kept the Comet's name to the fore with several further record achievements. Although K 5084 lingered unloved for a while following its short but eventful career with the RAF, Arthur Clouston, who was one of the pilots who had flown it whilst in military ownership, was determined that it should fly again as a civil machine. The Air Ministry had sold it to a scrap merchant and on 21st January 1937 this advertisement appeared in *Flight* under Aircraft and Engines for sale:

> 'DH Comet (actual winner of London — Melbourne race) fitted two Gipsy Six R engines, Ratier V.P. Airscrews. Undercarriage and Cowlings damaged. Offers Wanted. L. Lipton, 614 Old Ford Road, London E3. Phone: Advance 3345.'

Mr Lipton disposed of it for £250 to Frederick Tasker of East Barnet who sent the crumpled package to Essex Aero Ltd at Gravesend Airport; there Jack Cross, who had considerable experience in preparing aeroplanes for racing and record purposes, devoted his expertise to a major rebuild and he incorporated several marked modifications. Standard Gipsy Six II engines replaced the rather temperamental high-compression Six Rs and to these were fitted de Havilland's new variable pitch propellers. New less-tight-fitting cowlings were used and these had larger air intakes to keep the engines at lower operating temperatures. By use of a new design of heavier Dunlop tyres, the maximum permitted take-off weight was increased to 5,720 lbs, showing an improvement of 170 lbs.

In this form G-ACSS re-emerged on the civil scene. Painted in a blue and grey scheme and renamed *The Orphan,* the newly-refurbished Comet was entered in a race to be staged in August 1937 from New York to Paris to mark the tenth anniversary of the famous solo Atlantic crossing by Charles Lindbergh. There were twenty-one published contestants, but shortly before the event the United States Government expressed serious doubts about the safety of so many pilots and aircraft crossing so much water and the U.S. Department of Commerce banned the race.

Almost immediately an alternative route was announced and this was to be from Marseilles (Istres) to Damascus, to Paris (Le Bourget). The number of starters had dwindled to thirteen and G-ACSS was the only British machine to be entered. Carrying the black racing number G-16 in a large white rectangle on the nose (a most unusual position, for normally such numbers are painted on the vertical tail surfaces) *The Orphan,* flown by Flying Officer Clouston and Flight Lieutenant George Nelson, averaged 196 mph over a route of 3,850 miles in an elapsed time of 19 hours and 41 minutes. The Comet made the fourth fastest time to an Italian Savoia Marchetti SM 79 that achieved 219 mph for the course. The race was flown on 20th and 21st August 1937.

That is all to be found in the history books, but any flight of this nature must make more worthwhile reading than the bare statistics that it produced. To appreciate the strains imposed on the aircraft and its occupants, no such tale can be more convincingly impressive than these selected extracts from a report written by Clouston shortly after the event:

> 'During the France — Syria — France Air Race from Istres (Marseilles) to Damascus (French Syria) and back to Paris, a storm belt of 450 miles was flown through over the Swiss Alps at 18,700 feet where severe icing conditions were encountered. This storm belt extended from the Mediterranean commencing about 150 miles due East of Venice to within 180 miles of Paris. With the exception of this storm area the entire 3,849 mile course was flown in good weather.
>
> The storm was entered at 7,000 feet on a Great Circle course of 302°. The course Damascus — Paris direct takes one over the middle of the Swiss Alps which extend for over 450 miles and reach a height of 16,000 feet in places.
>
> The majority of the 2,000 miles run home, Damascus — Paris, was flown just clear of the mountains at 6,000 — 10,000 feet in good weather, but when some 150 miles from the Swiss Alps the above mentioned storm belt was entered at 7,000 feet and the Comet was climbed through it to 18,700 feet above 10/10 clouds in sunshine. The aircraft was climbed

G-ACSS flew under various names. Here, as 'The Orphan' and with the racing number G-16 displayed prominently, it is seen at Gravesend in August 1937 prior to the race from Istres to Damascus. Note the absence of discs on the spinners, as evidence that the Ratier propellers have been replaced by more modern controllable pitch units.

through the storm area slowly so as not to lose too much time in the Race, the climb from 7,000 feet to 18,700 feet taking nearly an hour. During the climb heavy rain, then hail and snow clouds, were passed through. The Comet took on about ¾″ to 1″ of ice during this climb.

The clouds flown through on the climb from 7,000 feet to 18,700 feet were very heavy rain clouds with terrific up and down currents and some lightning up to about 15,000 feet where the rain changed to heavy hail showers, then snow until 18,700 feet was reached above the storm in sunshine.

The Comet had collected about ¾″ to 1″ of ice on the leading edges of the tail plane, fin and main planes during the climb. Later on the clouds extended to probably 30,000 feet, 10/10 and were of surprisingly varied types. At times they were heavy cumulus that almost obscured the wing tips and housed a colossal amount of moisture that froze readily to the aircraft. At other times belts of heavy snow storms were passed through which did not freeze to the aircraft readily; then again we would pass through a heavy hail-stone type of cloud, and again through heavy rain, all at 18,700 feet — 19,000 feet and blind flying 10/10 clouds all the time. It was very disconcerting weather to fly through and the aircraft was pitched about in gigantic bumps for several minutes, then would settle down on a steady blind flying course as the clouds varied, only to be pitched about with force as the hail and snow belts were intermittently flown through.

The frontal portion of the Comet screen became iced over and completely restricted forward visibility. The entire screen is made of Rhodoid and during a period when the machine was spiralling down in the region of 13,000 feet a loud report, similar in sound to a shotgun report, rang out. As we were still in 10/10 cloud, knowing the mountain peaks were 16,000 feet, George Nelson and I naturally assumed we had hit the mountains but almost immediately our cockpits were filled with hail, snow and rain. The complete cockpit Rhodoid covering, extending over and around both cockpits, had blown up into small pieces and disappeared with the exception of a few small strips of Rhodoid attached to the ribs forming the cockpit housing, which were left flapping in the air flow, plus a small frontal piece.

This extraordinary bursting of the cockpit covering I can only assume was caused by the great change in temperature as the Comet lost height quickly with failing engines from 18,700 feet to 13,000 feet. At 13,000 feet we found ourselves in a break in the clouds, with no sky or earth visible. It was obvious we were down a valley in the mountains as the mountain peaks were 16,000 feet in this area. This therefore probably caused considerable changes in the temperature. From then on until the ice areas were left both George Nelson and myself experienced a very uncomfortable journey sitting in the flying hail and snow at 200 mph which feels like hitting cricket balls. We then found ourselves at 12,800 feet circling around in a break in the clouds that might have closed up any moment and with no top, sides or bottom visible.

We could not proceed or return at this height as the mountains extended up to 16,000 feet all around, although not visible. We therefore had no option but to spiral up into the snow and hail, that had already sent us down, until above 16,000 feet, before continuing on a course for Paris. About 20 minutes was spent in climbing the Comet to 17,000 feet before setting a course for Paris, which was still in a 10/10 snow storm with occasional patches of heavy hail and rain.

The engines were running smoothly at about 2,000 revs. and continued to do so for about 12-15 minutes, when they both began to lose revs. slowly, running roughly until just a tick-over was obtained.

As the engines began to lose revs. I tried opening and shutting the throttles, then the altitude controls and finally the hot air intakes themselves, but no recovery or appreciable response was evident from the motors.

The Comet lost height in the snow storm to approximately 11,000 feet where warmer air was again met and the instruments once again fogged over inside the glasses. A long break in the clouds running North and South with still no sky or ground visible, was glided along. No sooner had this break been entered than the engines increased in revs., up to 2,100 and ran smoothly.

It was obvious that either my petrol tank vents were icing up or hot air controls at the carburettor were strained owing to being forced on when iced up when the engines first showed signs of failing.

As we were still in among the Alps below the highest peaks and ranges and surrounded by cloud, I had no option again but to stay in the cloud valley until quite sure my engines were functioning satisfactorily, then spiral up again into the ice and snow that had driven me down for the second time. Again I climbed the Comet to 17,000 feet and continued on course for Paris. The engines ran well for 15-20 minutes when they both dropped revs. slowly but evenly from 2,000 to just a tick over. The hot air controls at the carburettor were left on and the throttles full open. The Comet lost height to 8,500 feet and passed through a heavy snow, hail and eventually rain storm. At this height we found ourselves down in a valley in the Alps with clouds on either side but a rocky bottom was visible. The engines picked up slowly to 2,150 revs. and ran smoothly. I flew down the valley losing height to come out at 4,000 ft under 10/10 clouds that were clear of the mountains. The hot air controls were pushed off and the engines increased in revs. to 2,280 at full throttle. Occasional heavy rain storms were flown through but the engines continued to run smoothly until the finishing line was crossed at Le Bourget 1 hour 12 minutes later.

The experience gained on this flight leads to the following observations:-

(a) How and why the Gipsy VI Series II engines in the Comet failed repeatedly in icing conditions with the hot air controls at the carburettors fully on? My own suggestion is that either my petrol tanks vent pipe iced up or the hot air control was not closing fully owing to jammed or strained control leads caused by the excessive cold (the control levers in the cockpit could be operated to the fully "on" and "off" positions after several attempts) although on conclusion of the flight these controls were found to be functioning perfectly, both at the engines and cockpit.

(b) The colossal change in weather conditions flying through cloud over the Alps at 19,000 feet is interesting. While flying through a thick normal type of cumulus cloud in fairly calm air we would pass through a heavy hail storm, then rain, then again snow and back again probably to ordinary watery type cumulus cloud. This amazing change in the clouds I consider must be due to the varying wind velocities and air currents peculiar to mountainous areas. It is very disconcerting weather to fly through, as the bumps at times in some of the belts of hail and snow are terrific and there is no indication one is changing to rain, hail or snow, or *vice versa,* until such cloud has been entered.

When passing through the watery type of cumulus clouds ice formed quickly with rough jagged edges and with a clear glassy appearance. After passing through a heavy hail storm for several minutes at the same height, the jagged edge was knocked off the ice formation on the wings. This is a phenomenon I have never experienced on any de-icing flights over England.

I suggest some of the hail belts passed through housed larger hail stones than experienced in England. The leading edge of the Comet main plane and tail unit was worn to the bare wood which indicates the continual beating of large and heavy hail stones.

In conclusion this flight over the Swiss Alps through the aforesaid icing conditions is the worst flight I have ever experienced during my flying career. The severity of icing on this flight was not as heavy as many I have experienced on the Northrop over England, but the abrupt and severe changes in meteorological conditions found over the Alps seems to indicate every precaution should be taken if flying over high mountainous areas when icing conditions are known to exist.

To have to sit in the Comet with no parachute, with both engines failed to a tick over, and to have to spiral three times between the mountains in ice, hail and rain clouds during the course of an hour, with nothing but three reliable altimeters to indicate that *terra firma* is around one, is the most horrible feeling I have ever experienced.'

(Sgd) A. E. Clouston
Flying Officer, R.A.F.O.
September 1937

Although frequently the Comet is referred to as a racer, the DH 88 participated in relatively few normal closed-circuit speed events and clearly the characteristics were more finely matched to straight point-to-point records than to shorter-course races round the pylons. However, only a month after Clouston's Damascus flight, Ken Waller was aboard again with 'CSS as one of thirty-one entries for the 1937 race for the King's Cup. On 10th September a preliminary heat — or elimination round — was run on a pure speed basis with no allowance for handicaps. This started from Hatfield and the route via Cambridge, Skegness, York, Scarborough, Whitby, Newcastle, Edinburgh, Aberdeen, Glasgow, Portpatrick and Newtownards to Baldonnel, covered 786.6 miles. On the next day, the final was flown from Baldonnel to Hatfield via Belfast, Portpatrick, Carlisle, St. Bee's Head, Blackpool, Stoke-on-Trent, Desford and Cardiff. Although Waller averaged 214 mph, this gave him no better than twelfth place, in one of the very few events contained wholly within the British Isles in which the Comet competed.

Now G-ACSS was back in its element in the long-range business, with the Cape as the target. Clouston was at the helm again, with Mrs. Betty Kirby-Green as the second occupant, but for some reason the Royal Aero Club refused to accept her as the co-pilot and the official records state unfairly that she was a passenger. Both were wearing Burberry flying kit and for this record attempt the Comet carried the manufacturer's name. On 14th November *The Burberry* left Croydon, to cover the 7,200 miles to Capetown in 45 hours and 2 minutes and yet another record was among the Comet's credits. However, this was not enough for Clouston and Mrs. Kirby-Green, for on 18th November they were airborne again to arrive back at Croydon after 57 hours and 23 minutes.

On 14th November 1937, G-ACSS, now renamed 'The Burberry', was flown from Croydon by Flying Officer (later Air Commodore) Arthur Clouston and Mrs. Betty Kirby-Green to knock 3 days, 16 hours and 2 minutes off the record to the Cape and back.

As always, there is more to the tale than the formally recorded facts. For this flight to the Cape and back the Comet was hired from Mr. Tasker for £1,200 and the insurance cost at least £500. Altogether the direct expenses for the flight were more than £2,500 and Mrs. Kirby-Green undertook to raise this money if Clouston would make all the flight preparations. Shortly before they were due to go, Tasker appeared at Croydon and declared that the agreement was cancelled as he had been given £50 short of the promised figure. Then the last few pounds were raised by passing the hat round among the occupants of the airport restaurant, so the flight was on!

Following Clouston's experiences in the Damascus — Paris race, the two occupants of G-ACSS had agreed that they would take parachutes. However, only one was delivered and no others were available at Croydon, so they went without them. For navigational purposes the flight was planned to take place during a full moon and the first take-off from Croydon, at full load in the dark, was not the smoothest of departures. To use Clouston's own description, written sixteen years afterwards:

'It was impossible to see ahead, and my only guide to keeping a straight course along the flarepath was a sideways view at an angle of forty-five degrees to the line of flight. The rough, grass surface of the airfield was not making it any easier to hold the Comet straight and it seemed quite an interminable time before I could lift her tail.

Suddenly we hit a bigger than usual bump and swung off the flarepath heading straight for the houses. We scraped over the chimney pots much too low for my liking, but at least we were airborne; and I turned on course for Marseilles.

Betty was chattering away quite unconcerned, thrilled with everything. She had accepted the take-off as a matter of course. I was glad I had not a more experienced pilot with me.'

After 10 hours and 55 minutes in the air, G-ACSS arrived at Cairo (a record in itself) and following refuelling and a brief breakfast *The Burberry* and its crew were en route for Khartoum. Then there were problems, for the local representative of the petrol company refused to put any fuel aboard unless a permit was issued by the Khartoum authorities. This was not obtainable without detailed information about the route to be taken and the preparations to be made for dealing with emergencies. As the Government offices were closed, Clouston, infuriated but admitting temporary defeat, retired to bed and assumed that his lady co-pilot also had done so. However, in the middle of the night she appeared, with the permit, to wake Clouston and tell him that the aeroplane was being refuelled and they could leave as soon as they wished. Apparently, in a borrowed car, she had travelled from lead to lead until she found the required official at a friend's house.

At that time Khartoum had a civil airport (which the Comet used) and a Royal Air Force station, each adjoining the other and separated by a single wire fence. Normally the longest available run was 1200 yards and this seemed insufficient for a full-load take-off in the warm conditions, but the Service people obliged by removing the fence and laying a flarepath for 2000 yards across the sands of the two airfields combined. Apparently 'CSS was reluctant to become airborne even after that length of run and when it did stagger from the ground it seemed unwilling to climb. So, as Clouston described it 'I dare not climb. We were no higher than a camel's back. Praying that we would meet no obstruction, I furiously wound the thirty-three and a half turns of the wheel in the cockpit that folded up the undercarriage, and flew with the other hand. It took only two or three minutes, but it seemed like eternity before the undercarriage was up and I was able to climb. . . .'

An entry in the Comet's logbook 'rough weather near the equator' meant all that and much more. Again, to quote first-hand experience from the cockpit:

> 'We were being thrown about unmercifully, and maps were fluttering up and down between the floor and the hood of the cockpit. The rain was like a solid wall of water and leaked into the cockpit. Everything I touched was saturated.
>
> As we crossed the equator the rain turned to hail. The force with which the hailstones clattered against the windscreen and canopy worried me as much as the fight I was having to keep the Comet on an even keel. I remember thinking that when we got back, no one was ever going to believe the story of a hailstorm on the equator.'

The next landing, at Broken Hill, near Lake Victoria, was ahead of schedule and the crew were told to move the machine from where they had parked, as the space was reserved for a record-breaker that would be arriving later and would need immediate refuelling! While the contestants were having breakfast, a signal was received to say that the Comet was on its way. . . .

The flight from Broken Hill could have been completed in one leg to Capetown, but Clouston elected to make a landing and a brief stop at Johannesburg just to break a record to that point previously set by C.W.A. Scott. Both occupants remained in the cockpit while the aeroplane was

Final fling: G-ACSS, in new guise as 'Australian Anniversary', at Croydon on 26th March 1938, after Arthur Clouston and Victor Ricketts had flown the 26,500 miles from England to New Zealand and back in 10 days and 21 hours. Four separate point-to-point records were set-up on the journey. This was the last achievement by a DH 88 Comet.

refuelled and they, too, received light sustenance in sandwiches and cold drinks. They were airborne again only 27 minutes after touch-down. Allegedly at Capetown one hundred thousand people broke down the fences and thronged round the Comet before the end of the landing-run and, to avoid accidents, Clouston switched-off the engines while the machine was still on the roll.

The route back was almost identical with that of the outward flight, but with an additional stop at Kisumu and an unplanned landing at Marseilles for fuel. At one stage the Comet was flying at 12,000 feet in an attempt to overfly the worst weather but then severe icing caused the aircraft to lose height steadily. Fatigue played its part on this return stretch and Clouston make five attempts to land at Cairo before pulling-off a successful arrival.

The out-and-back flight, which totalled 14,690 miles, took only 5 days 17 hours and 28 minutes and this knocked more than three and a half days from the previous record.

This was not the end either for G-ACSS or for Clouston, who was joined by Victor Ricketts to attempt a record flight from England to New Zealand. Now named *Australian Anniversary,* the Comet and its crew departed from Gravesend on 6th February 1938, but due to unforecast headwinds they made an unscheduled landing for fuel at Adana in Turkey. Both pilots were arrested, but subsequently released. Unfortunately the aerodrome was waterlogged and take-off was impossible, so the record attempt was abandoned. The aircraft was moved to a flint roadway so that it could be flown back to England, but due to a tyre burst (probably resulting from the flint surface) the undercarriage was damaged on landing at Nicosia. Temporary repairs covered the flight home to England via Brindisi. Delayed but not dispirited, Clouston and Ricketts planned to try again.

Following extensive undercarriage repairs by Jack Cross, on 15th March the famous Comet left Gravesend for a further attempt, this time successfully, to arrive at Blenheim, New Zealand in a record of 4 days 8 hours and 7 minutes. After only 16 hours on the ground, Comet and crew were in the air again for the flight home, which they achieved without undue incident. The total time away from England was 10 days and 21 hours and the overall distance covered was 26,500 miles to make this the first-ever direct flight from England to New Zealand and back. In addition to this achievement, the en route speeds brought several point-to-point records, including London to Sydney in 80 hours 56 minutes at an average of 130.3 mph.

Flights of this magnitude call for very positive planning and practice in the skills of navigation. Today we might know very little about the way in which the Comet pilots coped with this part of the total task, but fortunately Arthur Clouston wrote these words in *Flight* dated 7th April 1938:

> 'Now for navigation itself. Working out the course from A to B, I take the direct bearing and mean variation, and set that course off on the compass. I have never yet allowed for wind drift in the Comet, not even when I have known there to be a strong cross-wind blowing. It has been my experience that meteorological reports of weather for 2,000 mile stretches at 7,000 ft — the Comet's operating height — cannot be trusted.

The damaged G-ACSS – now named 'Australian Anniversary' – following a tyre burst on landing at Nicosia in February 1938. This was on the return flight to England after the first, unsuccessful, attempt to reach New Zealand.

(Vintage Aircraft magazine).

The upper winds have always turned out to be different from the way they should have been. It was in the Johannesburg Race, 1936, that I discovered drift can be ignored in machines that cruise at around the 200 mph mark over several thousand miles. In twelve hours' flying cross-winds will practically always cancel themselves out.

When Ricketts and I started away on our first unfortunate crack at the New Zealand trip in February we found ourselves at one point over Europe 40 miles north of our course. We just kept to our compass course, and three hours later, after blind flying, we were 30 miles south of it. I never correct such flying for drift — I wait for the correcting wind to turn up a little farther on. Whenever possible, I check up my position on definite landmarks like rivers, mountains and islands, and keep marking my position on my maps. At night, unless there is a strong moon, there is scarcely anything to be seen. Over 2,000 miles probably half of the course is under cloud. And, in any case, crossing a country like India by night brings up very little indeed to check your position on. Through the dark hours I just sit on my compass course, speed, height and time. I try whenever possible to schedule my refuelling stop for daylight. It is madness to go off for ten hours' flying and then expect to find an aerodrome in the middle of a strange country like Australia, Africa or India in the dark.

One should be able to owe accuracy in navigating to one's own confidence in course-keeping. Sometimes I have wanted to think my compass was out by 20 or 30 degrees when mountains, or snow peaks, or islands sticking out of the sea have given me a cross-bearing a bit different from that which I had expected. But in every case I have stuck to my original course, and the good old P.4 has turned out to be right. I would trust a properly swung and adjusted compass for a month.

When I get a definite check by landmark (confirmed by a second check) I work out ground speed and keep on working it out. I keep applying the result of my little arithmetical sums to the distance still to go. Flying up to 1,600 miles across open sea, as we did to New Zealand, we found that wind direction as shown by seaspray was frequently entirely different at our 7,000 ft. We always kept to that height, no matter what storms lay in front. I only fly round cloud patches over land, and when I am fairly certain of where I am. If you begin to make five-mile detours over the sea your landfall after 1,000 miles may be very different from the one neatly drawn on your map.'

The last public appearance by G-ACSS was as an entry in the King's Cup Air Race for 1938, but taking-off from Hatfield on 1st July (the day before the event) the Comet suffered from a tip-stall and the port wing and aileron were damaged. Among those who saw the incident were Alex Henshaw, Jack Cross and Ron Paine. The last of these, then an up-and-coming young engineer, was very interested in the repair scheme, for which A. E. Hagg signed drawings to show butt jointing instead of scarfing; this was because the wingtips needed to endure more harsh man-handling loads than flying loads. Albert Grant, the

DH Chief Service Engineer, explained how the drawings and technical records were kept in the Service Department and it is interesting to note that 45 years later, for the restoration programme described in chapter 9, Ron Paine was able to obtain these records (No. R 900) from Chester within 48 hours! The airframe logbook shows that five weeks after the accident, when repairs had been completed, two test flights were carried out by Geoffrey de Havilland; these are the only entries showing that he flew this aircraft.

Then came major work in preparation for another planned flight to Australia. Both engines were removed and sent to Stag Lane for complete overhaul, while the propellers were modified to take constant speed units. (A constant-speed propeller is one in which the rpm remain constant, within a certain range, when the throttle is opened or closed or when the aircraft is dived or climbed. Without this facility the revolutions of the propeller will change with alterations in the aircraft's throttle setting, attitude or speed).

On 5th November 1938 the Certificate of Airworthiness was renewed, but a logbook entry for the following day shows that the propellers were removed again and replaced. G-ACSS did not fly again and the final record in any of the documents shows that it was requisitioned by the Royal Air Force at the surprisingly late date of 14th September 1943, shortly after which it was sold to the de Havilland Aircraft Company. The total recorded flying time by then was 510 hours and 3 minutes.

The Australian flight, of course, had not taken place. Instead, G-ACSS spent the first part of World War II hangared at Gravesend, but when the space was needed for military aircraft the most famous Comet of all was pushed outside to spend several ignoble years under a camouflaged tarpaulin sheet. The elements made their marks and naturally serious deterioration occurred, but G-ACSS survived the war as a complete aeroplane.

That accounts for the lives — and fates — of all but one of the DH 88s. Three were known to have been destroyed and one remained partially intact, but at this stage the state of the Portuguese DH 88 remained untraced. There might have been other Comets to consider today, for in mid-1935 there were plans to build several for a rich businessman; also, at about the same time, the type was being studied seriously for use on a secret high-speed bombing mission. Records in the archives at Hatfield show that this version would have carried a 1000 lb bomb and be armed with a pair of Lewis guns. Yet another proposal was for a Comet with higher tankage to hold an extra 83 gallons, giving a range of 3800 miles, but none of these ideas materialised.

Although only five Comets were built, the design provided considerable scope for the development and expansion of the techniques used in its construction. The key people within the de Havilland company realised that as their small two-seat monoplane had competed in the MacRobertson race against established American airliners, Britain should modernise its products for the commercial field and understandably the result should be able to boast a DH tag. Therefore this letter was sent on 8th November 1934 to the Air Ministry:

'Although we, as the designers and manufacturers of the Comet which won the recent England-Australia Air Race, have received many hundreds of congratulatory messages and much favourable comment in the World's Press, it is unquestionable that the fine performance of the American Douglas and Boeing machines — standard models as used on the air mail lines of the United States — demonstrated how far this country lags behind in the use of high-speed commercial aircraft.

'The Comets, all of which established records during or after the contest, although produced especially to conform to the conditions of the race, have practically no future as a production or saleable type. In designing the Comet, we had to have prominently in mind the speeds achieved by American commercial aircraft, and it was evident that our best chance of success against them lay in providing a machine of even higher speed combined with greater range. These attributes, however, in the form of the Comet, do not at present meet any demand of existing Empire Air Routes.

'It will not be out of place to mention the heavy cost shouldered by this company over the production of the Comets and the necessary race organization involved. Three machines were built, together with a reserve for spare parts, likewise ten special engines and spares, also twelve experimental variable-pitch airscrews. Suffice it to say that the purely nominal prices paid by the entrants have recovered little more than 25 per cent of the total direct cost in which we have been involved. You will understand, therefore, how immensely important it is to us to derive some benefit from the race.

'We believe that this company has pursued the right policy up to the present time in producing types of commercial aircraft which can be operated economically with the minimum of artificial financial aid, and it will, in the ordinary course of events, follow on the same lines in the future.

'It is, however, strikingly apparent to us that immediate steps must be taken in this country to produce aircraft of better performance and of greater commercial merit than the successful foreign machines in the MacRobertson Race — and their inevitable developments.

'We are confident that, as a result of our experience in the development of the Comet combined with the wide knowledge we have acquired in the production of commercial types during the past fifteen years, we could produce a commercial aeroplane which would acquit itself with the products of any nation.

'The production of such an aircraft and of the engines required for it, involves greater cost than we, as a company, are justified in shouldering until we see that sufficient need will exist for it to enable us to recover our expenses. We interpret the signs of the times as indicating that demand will probably need to be stimulated by some form of Government assistance if our national prestige is to be maintained. We therefore venture to enquire whether some form of financial assistance could be extended to us, so that we may proceed with the design of an aircraft and engine suitable for offering in the world's markets, on our own air routes, and in competition with foreign lines; and which may be at least as good as those likely to be forthcoming from America in two years' time.

'It will be unnecessary in this communication to refer in greater detail to our proposals, but we should greatly appreciate an opportunity of discussing the matter with the appropriate authorities if the principle can be agreed upon.'

This appears to have been accepted favourably, but despite several discussions between the company and the Ministry, after nearly four months no reply had been received and on 1st March 1935 this "chaser" was despatched:

Prototype High-Speed Transport Aircraft

'Four months of valuable time have elapsed since the significant results of the Australia Race when, foreseeing the clear lines of urgent development, we sought your assistance with proposals for the development of an advanced commercial aircraft and engines, and your decision has now become a matter of urgent importance.

'We believe that failure to take account of American progress and to act promptly and effectively to recover our backwardness in commercial aviation equipment will have far-reaching consequences, and the overseas markets, which are building up a network of world air communications and can provide an outlet for British trade of all kinds, will be lost unless we can outpace our American competitors who, taking every advantage of a progressive home market forced by liberal State assistance are making technical progress which, it must be admitted, is rapidly widening the gap between us.

'One by one the big European air transport companies are turning to American equipment, and the same tendency can be observed in South America, in the East, and even in the British Dominions; further, we would hazard the guess that military flying equipment will follow as an inevitable corollary.

'We are perfectly confident that we can produce aircraft more acceptable to the world's operators than American types; but we do not feel justified in undertaking the initial development work without some show of official interest and some direct financial assistance.

'We venture to suggest, therefore, that the lapse of time and the importance of the subject justify us in expecting an early answer either favourable or otherwise; because until we know whether our proposals are of interest to our Government or not, we are unable to formulate our future policy.'

This time, action was quick, but frustrating, in the form of the following reply dated 8th March, 1935:

'With reference to your letter of the 1st instant and to earlier correspondence relating to the development of an advanced high-speed commercial aircraft and engines, I write to inform you that the best means of securing this development is still under consideration.

'It has, however, been decided that it will not be possible for the Department to place a non-competitive order with a single firm for the construction of a prototype, and that your particular proposal cannot, therefore, be entertained.

'I regret that it has not been practicable to give you an earlier decision in the matter.'

Despite this, the company continued negotiations with both the Air Ministry and Imperial Airways, as a result of which the DH 91 Albatross appeared. This was a sleak, low-wing four-engined airliner with a cruising speed in excess of 200 mph, with capacity for 22 passengers and a crew of four; the main point here, though, is that the DH 91's one-piece wing with a span of 105 feet was constructed in the same manner as that of the DH 88, with two layers of diagonally-placed spruce planking providing much of the strength. The fuselage was built with laminations of cedar ply and a layer of balsa wood in between.

This was not the end of the Comet's contribution to further achievements. In 1938 the de Havilland company was aware of the likelihood of war and after achieving such success in speed terms through the aerodynamically clean lines of both the DH 88 and the DH 91, designers' minds turned to the possibility of a bomber version of the Albatross. Progressively, though, the concept changed and the eventual outcome was the DH 98 Mosquito. The story of this famous wooden wonder has been related too many times to justify repetition here, except to mention the constructional method: the fuselage followed the practices used on the Albatross and the wings used the same system as that used on the airliner and, before that, on the Comet.

Without the Comet there would have been no Albatross; and without either the Comet or the Albatross almost certainly there would have been no Mosquito. Although the aim behind the DH 88 Comet was wholly civil and peace-loving, there is !ittle doubt that without it Britain would have lacked the most versatile of all the operational aircraft that served in World War II. This is not the place to consider what effect its absence might have had on the results, but let us record for all time that credit for the Mosquito's existence must be placed squarely on the list of the Comet's claims to fame.

Chapter 7
The Comet as a flying machine

Many people have written many thousands of words about the Comet: the reason for its origin and existence, details of design and construction, the success in the MacRobertson race and afterwards and its effect on the future of world-wide commercial aviation. Yet little has been recorded about its characteristics and qualities as an aeroplane to fly.

In the thirties, not only was flight test reporting relatively unrefined, but the nature and extent of the tests demanded by the airworthiness authority left gaps in the available knowledge about a type's performance or — more significantly — its shortcomings. Today a routine flight test for the *renewal* of a Certificate of Airworthiness, even on a simple single-engined light aeroplane, demands the use of a report form that occupies eleven pages. Admittedly two of these pages may not apply to a machine that lacks certain equipment — for example if radio is not fitted — but the performance statistics are required for filing by the Airworthiness Division of the Civil Aviation Authority so that each individual aeroplane's behaviour can be monitored throughout its flying life. Each type has a published figure for the rate of climb and a result close to this figure (adjusted for temperature) must be achieved on the test. A tired engine, a pitted propeller, a battered airframe, a load of odd items that may have been added at various times and other changes to the machine's overall health will be revealed on these regularly repeated tests and if the performance falls markedly below the norm, the CAA — or the owner if he is so inclined — can call for an investigation.

To achieve this life history there must be a starting point. On the form used for C. of A. renewal tests various statistics must be recorded and these can be checked as the aeroplane's age increases. Apart from direct comparisons with recorded figures, such as climb performance, oil pressures and maximum obtainable rpm, pure flying characteristics can change and these can give cause for more detailed examination: a machine may have tended to drop the port wing at the stall and after several years the opposite wing goes down. Clearly one stall is inadequate as a test, for various outside influences can affect the aircraft's behaviour, but if six successive stalls consistently produce a changed result, the time has come to seek the reason. With a wooden aeroplane in particular, changed temperatures and degrees of dampness can affect the airframe and some minor warping can alter its conduct in the air.

We are all aware of the frantic rush in which the Comet was invented, designed and flown, but all the figures and other information required for a C. of A. renewal in the eighties can be acquired in one flight or, occasionally, in two trips, so it is not difficult to obtain. Sometimes the weather may be suitable for the normal checks to be made at fairly low level, but may prevent manoeuvring at height, especially if the type is aerobatic and a spin check is required. Another reason for a pilot not completing a full flight test is turbulence. The schedule requires the aircraft to be dived to a figure approaching its designed limiting speed, but only the *makers* took the Comet to 300 mph; this would be to check that no flexing occurs that might change the machine's characteristics and to ensure that no flutter develops on the flying controls. No responsible pilot would carry out this check unless the air is reasonably smooth and sometimes a delay is unavoidable whilst awaiting better conditions.

Whilst we must all agree that safety makes sense, far too many aeroplane accidents continue to occur and we should check whether these 'developments' have created any *real* improvement since the days of the Comet's introduction. In 1983 the accident rate among British-registered light aircraft was the worst for seventeen years.

I mention the modern procedure as a direct comparison with the limited requirements of 1934. Although time was tight, more performance statistics could have been noted and recorded, yet the formal type test for the initial C. of A. clearance provides very little figurative information. At that time all official trials were carried out by the Aeroplane and Armament Experimental Establishment at Martlesham Heath. Although primarily a station with a Service core, both military and civil aeroplanes underwent proving tests through this Suffolk base. The best way to show readers the extent of the information either sought or obtained in 1934 is to reproduce the official report in total, for which I am indebted to the Airworthiness Division of the Civil Aviation Authority:

C. of A. REPORT NO. M 656/CA
October 1934

AEROPLANE AND ARMAMENT EXPERIMENTAL ESTABLISHMENT

MARTLESHAM HEATH
Suffolk.

Light Load Handling Trials of de Havilland "Comet" E.1., — 2 Gipsy VI.R.

Authority:	328576/34/R.D.A.6.	
Constructor:	The de Havilland Aircraft Co. Ltd.,	
Engines:	Gipsy VI.R.	International r.p.m. 2350
		Maximum permissible r.p.m. 2400
Airscrew:	Ratier — variable pitch	

LOADING AND CENTRE OF GRAVITY POSITION

The trials were made at Messrs. de Havilland's aerodrome at Hatfield, the loading of the aircraft being certified by the Resident Technical Officer.

The details of the two loadings at which the aircraft was flown are as follows:-

Case (1) **C.G. normal**

Weight empty	2930 lb.
Pilot and parachute	200 lb.
Fuel — front tank (50 galls)	385 lb.
centre tank (50 galls)	385 lb.
Oil (14 galls)	136 lb.
Total Flying Weight	4036 lb.

C.G. position:- 31.8 inches aft of the leading edge of the main plane measured at a distance of 24.75 inches from the plane of symmetry of the aircraft.

Case (11) **C.G. aft**

Weight empty	2930 lb.
Pilot and parachute	200 lb.
Fuel — front tank (47 galls)	362 lb.
centre tank (50 galls)	385 lb.
rear tank (20 galls)	154 lb.
Oil (14 galls)	136 lb.
Total Flying Weight	4167 lb.

C.G. position:- 36.5 inches aft of the leading edge of the main plane measured at a distance of 24.75 inches from the plane of symmetry of the aircraft.

Handling.

With the aircraft loaded as defined above it was flown over the whole speed range both with engines on and off.

Controls.

The elevators are light and effective at all speeds. The ailerons are very light and effective at all speeds.

The rudder was light and effective at all speeds.

Stability.

The aircraft is unstable laterally, and stable directionally and longitudinally. The stall occurs at approximately 84 mph (A.S.I.) with the flaps up and 74 mph (A.S.I.) with the flaps down. The stalling speed was difficult to judge as the stall occurs slowly, the aircraft shaking at speeds about 8 mph above the stall. The nose drops slowly and either wing may drop but control can be quickly regained.

The aircraft takes off in a normal manner. With the airscrews set at their coarse pitch, it is only just possible to taxy the aircraft and any attempt to take off with the airscrews at coarse pitch would involve a very long run.

The approach glide is flat, even with the flaps down but the aircraft can be brought in under full control, the landing being straightforward except that care has to be exercised owing to the high landing speed, poor view and light elevators.

The undercarriage absorbs the shocks well, but there is insufficient damping. The wide track undercarriage results in a very quick rock from side to side whilst moving over rough ground. The retracting mechanism is simple to operate but slow in action, 14 turns of a wheel being required to wind the wheels into either the up or down position. A red light interconnected with the throttle gives warning if the throttle is shut when the wheels are not fully down.

In level flight the view is good except forwards and downwards. During take-off and during the approach and landing the view is very restricted owing to the long nose and flat angle of glide.

General.

There is no change of trim engine-on to engine-off but when the flaps are lowered considerable tail heaviness is experienced. This is however easily counteracted by the spring loading device fitted to the elevators. Ample tail range and control is available for the range of centre of gravity.

With one engine throttled the aircraft handles well, only slight pressure on the rudder being necessary to maintain a straight course. At the light loads at which the tests were made and with the airscrew set at fine pitch the aircraft climbs at 5000 feet; with the airscrew set at coarse pitch it will maintain height at 5000 feet.

The weather was bumpy at the time that the tests were made. Slight flexing of the wing tips was noticed during bumps.

These two flights, of 35 and 25 minutes duration respectively, were carried out by Flight Lieutenant J. N. Boothman, whose logbooks I have examined; he conducted the tests at Hatfield on 5th October, flying from and back to Martlesham in an Avro Tutor (K 3189). That is the total extent of the *official* flying for the Comet's type certification, but a subsequent maker's test flight is relevant here and this is the report on file:

D. H. COMET
GENERAL HANDLING TRIALS.
at 5100 lbs.

The general handling qualities of this aircraft are normal and straightforward and all the controls are positive and light.

1) **Elevators.** Since the original flights the elevators have been lower geared as they were too sensitive landing and a heavy-handed pilot would be inclined to balloon the machine when flattening out. They are powerful enough to overcome any C.G. position without use of the trimming gear.

2) **Ailerons.** These are exceptionally powerful and light and have been stiffened by use of springs to make the feel on the control lever heavier. They are effective down to, but not past, the stall.

3) **Rudder.** The rudder control is ample at all speeds and has enough area to cope with one engine flight even without use of the bias gear.

Stability. At the forward limit the aircraft is just stable (i.e. 31.8 ins. behind leading edge) 0.277 A.M.C.

At the aft limit (36.5 ins. behind leading edge) 0.348 A.M.C. it is unstable.

The machine has been taken off and landed on both these limits, and the brakes used in the normal way.

The aircraft has been dived to 300 mph (A.S.I.) on the aft limit and is exceptionally steady. There is no sign of flutter even with coarse use of ailerons. The tail unit is vibrationless and visual observation can detect no movement of any kind.

Flaps. The flap gear is extremely effective both in coarsening the angle of glide and in reducing stalling speed. The load on the pilot's hand is fairly heavy but this is not at all worrying as it can be used in four stages.

The use of the flap makes the aircraft tail heavy.

Although, of course, details of performance are recorded on the maker's test reports (of which many pages remain in the aerodynamics archives at Hatfield), surprisingly not even engine readings are recorded on the clearance flights for the C. of A. Certainly this was primarily a test on the aircraft and not on its power units, but the high-compression Gipsy Six Rs were liable to be temperamental and their performance on the Comet could be very different from the figures that had been experienced a few months previously on the relatively low-speed Leopard Moth. Temperature and pressure readings could have revealed much more than just the positions of the pointers on the gauges.

On the flying performance side, almost the only statistical information provided were the 'clean' and 'dirty' (i.e. wheels and flaps up and down) stalling speeds. These, of course, form the basis on which many other facts are deduced, but alone are they sufficient? The C. of A. report omits any reference to level speeds obtained at specific power settings and the rate of climb is not recorded.

Although both pilots of G-ACSS in the Australia race put pen to paper on many occasions, mainly in the form of newspaper reports, neither wrote anything about the aircraft's flying qualities. In Scott's book, which describes many of his earlier experiences in both Service and civil aviation, the only points of pertinence to the Comet's characteristics were immediately after his first flight at Hatfield, when he reported: 'All the time I was worried about the descent, because these Comets have such a terrific landing speed. However, we managed it perfectly'; and after crossing the Timor Sea he wrote 'flying the Comet on that one engine was a terrific strain, I assure you'. There is no more to be found.

As far as I can discover, apart from de Havillands' own General Instructions to Pilots, issued on 9th October 1934, more than two years passed before any handling information was written and published. Only one pilot is known to have gone to any depth of detail and fortunately he seems to have completed almost as much writing as flying. The person concerned was Flying Officer (later Air Commodore) A. E. Clouston, who died on 1st January 1984, but in the late thirties he provided several very useful reports about flying the Comet in a wide range of places and conditions. In the previous chapter you will have read about some of his exploits, in the course of which he gathered a significant number of point-to-point speed records.

Where pure handling is concerned, one of Clouston's main reports was written early in 1937 and was based on long-distance flights in G-ACSS, by then known as *Australian Anniversary.* The high-compression Gipsy Six Rs had been replaced by standard Gipsy Six IIs and the Ratier propellers had given way to more acceptable Hamilton variable pitch units controllable from the cockpit. His description is interesting:

> 'Landing the Comet is a matter to be dealt with in sequence. I like to arrive over my aerodrome and make four or five circuits before I land. It is not an easy aeroplane to see out of, and I like to have everything under control and observation.
>
> First the engines are cut down to about one-third throttle, dropping speed down from 200 to about 130 mph. Then the undercarriage is wound down. It takes 14½turns of the handwheel to get the undercart extended, and the wheel can only be turned about a tenth of a revolution with each movement. It takes four or five minutes to do the job calmly and efficiently.
>
> Next the airscrews are put into fine pitch, speed cut down again to 110 mph and the flaps brought on. They are intensely stiff, and take just about all your strength to bring down. Then round to land in a very wide approach turn with throttles half open. I take her round to have at least a mile in which to fly in to land at hot, sandy aerodromes.
>
> It is my opinion that one should on no account try to put the Comet down in a pretty three-point landing. It is the only aeroplane I know that must NOT be landed that way. It must always be a slightly "wheel" landing. If a three-pointer is tried, the Comet won't sit down on three points; she will just balloon off into space again.
>
> Landing is not easy, but taking-off fully laden from tropical aerodromes is a major problem. The machine is excessively tail heavy and very prone to swing. If it develops a good swing you catch it early with the motors and, if need be, the brakes as well. The fact that you can see nothing ahead adds to the difficulties.
>
> The engines are opened to half-throttle to allow them to start turning evenly and get the machine rolling nicely, and then increased to wide open. Partial swings can be corrected, if you catch them in time, by juggling the throttles only. The brake lever is left a quarter on. The brakes themselves are not working in this position unless the rudder bar is brought into play, when they work as another check on swing.
>
> The tail will show no sign of coming up for two or three hundred yards,

as full-load acceleration is so slow. But once the tail is lifted the Comet, as clean as a knife, has acceleration of a positively amazing order if it is held in flying position.

In spite of the apparent speed, it has to be held down the full length of the runway. If you try to lift it off before it is ready it will simply sit back again on its tail and slither — on into the hedge.

There is no normal sensation that you are about to fly as the Comet approaches the boundary. You must allow it to run and run until just before the hedge, when the stick is eased back. Then it will have attained about 90-95 mph and will fly off soggily. After the first upward climb it must be eased down slightly, and the airscrews put into coarse pitch after the undercarriage has passed the half-up position during retraction. And so up to 7,000 feet in a long, gentle climb at 140 mph on the clock.

Probably our most moving take-off was from Charleville, homeward bound. We had tanked up to 200 gallons for the hop across 1,700 miles of featureless Australian desert to Darwin, when closer inspection by car of the 830-yard runway caused me to order 40 gallons to be pumped out again.

Normally I must have 1,000 yards for safety, and in tropical heat something like 1,700 yards. With the 40 gallons removed we just cleared the hedge, after one second's suspense when the machine sank a few feet after being pulled off in the last five yards of available space.'

The last recorded flight of a DH 88 was on 4th August 1938. Anyone who was only twenty-five years of age then would be in his seventies today, so first-hand experience is not easy to find, although I have been very fortunate to be able to speak to Hugh Buckingham, who carried out some early Comet flights; and to Richard Clarkson, who was the test observer with Hubert Broad on much of the Comet's development programme. Therefore I hope that a few explanatory comments will be acceptable and will not be considered presumptuous.

Clouston stresses the Comet's tendency to swing on take-off and mentions correcting this by juggling the throttles. At that time and for several years afterwards this was the standard practice, but later, with the introduction of increasingly powerful aircraft, a more straightforward procedure was adopted: after lining up and running for a few yards to straighten the tailwheel, the pilot led with one throttle slightly ahead of the other until the tail was up and the rudder was fully effective in the slipstream; then both levers could be opened together steadily but fully or taken to the normal take-off stops.

To elaborate on the problem, the direction of the swing depends on the direction of rotation of the engines and therefore the propellers. If the propellers revolve anti-clockwise when viewed from the cockpit, as on all Gipsy engines and therefore on the Comet, the aircraft will tend to swing to the right. To counteract this the right, or starboard, throttle will need to be opened slightly ahead of the other. This general tendency to swing exists throughout the take-off run as power is applied and this is correctable with rudder only in the later stages, but another aggravation to the unwary pilot is an inclination for the aeroplane to change heading due to the gyroscopic effect when the tail comes up from the three-point attitude. Even low-powered tail-down twins such as the Miles Gemini with its pair of Cirrus Minor engines of only 100 hp apiece need the throttle-leading treatment, especially on a hard surface where there is no natural braking tendency, but the more powerful types require serious devotion to the task in order to stay straight. A crosswind from the 'wrong' side aggravates the swing.

Clouston's comments about the need for wheel landings and to avoid pretty three-pointers is based on the Comet's tendency to balloon, which is a quality that might be expected of such a clean aeroplane, but he does not mention the type's inclination to drop a wing on the approach to the stall. Aeroplanes with pointed wings have this habit and the de Havilland enterprise had considerable experience of the problem with some of their designs. The DH 89 Rapide could play tricks if landed on three-points in even a modest cross-wind, while the original DH 87 Hornet Moth proved to be too "hot" for the private owners for whom it was intended; so tricky, in fact, that all DH 87s on the British register were recalled for makers' modifications to incorporate square tips, which resulted in much more docile low-speed characteristics. Clearly the Comet suffered from this wing-drop tendency and Hubert Broad exclaimed this after the earliest flights; also Richard Clarkson has told me tales of both violent wingdrops and tendencies to swing on landing, so there were two good reasons for keeping-up the speed on the approach and for 'wheeling' it onto the ground. He referred to one landing that ended in a particularly vicious swing, with the aircraft coming to rest on the wrong side of the aerodrome. Broad switched off both engines and said to Clarkson: 'Come on. Get out. If it behaves like that they can come and fetch it back'.

Perhaps now we may delve a little further in the Comet's flying history. During the MacRobertson race in October 1934, when C. W. A. Scott was approaching to land at Singapore, most reports state that he was doing so in a downwind direction while others indicate that the trouble was a crosswind. Whilst fifty years later we are unlikely to discover the truth, we should consider the possibility that he was contending with both problems at the same time.

Landing in a crosswind from behind can be one of the trickiest situations for a pilot to handle. An aeroplane depends entirely on the flow of air past the flying control surfaces for these controls to have any effect. Towards the end of a landing run into wind the controls become progressively less effective, so increasingly coarse movements are needed to produce any result. When landing into a wind of, say 20 mph, that speed of air remains flowing past the surfaces even when the aeroplane has stopped. Also, by landing towards the wind the speed over the ground at touch-down is reduced by that wind strength and therefore the subsequent run is shorter: hence the two main reasons for the value — and with some aircraft, the necessity — of landing into wind.

A crosswind may create problems at the initial touch-down stage because of the need to counteract the aircraft's tendency to drift sideways over the ground. A bigger difficulty, however, occurs later for the reasons just described. Because of the large vertical keel surface toward the rear, an aeroplane acts like a weathervane and seeks to face into the wind, just as it heads straight into the airflow (equivalent of wind) when in normal flight. Therefore, if during landing the wind over the ground is from one side, the aircraft will tend to swing into the direction from which that wind is blowing. At low speed, full rudder may be insufficient to counteract the swing. In the worst case, if such a swerve develops, it may become wholly uncontrollable; this is because the centre of gravity of a tail-down machine is behind the main wheels and this weight will wish to continue ahead of the part of the aircraft in front of the wheels. The resulting, uncontrollable, swing is known as a ground loop. These problems are virtually non-existent on a modern aeroplane with a tricycle undercarriage, but they would have been prevalent with the Comet.

Now let us go one stage further. We have a crosswind to handle, but if it is from behind the direction of approach all the previous problems arise, with an added hazard at the end of the landing run. If a machine finishes its run in calm conditions, there is no airflow over the controls (notably the rudder) at the point of stopping, but if the wind is from behind, at this stage there is a negative airflow (i.e. hitting the control surface from the rear) with a resulting total loss of control over the aeroplane's antics. This may well have been the position faced by Scott on his arrival at Singapore, but all reports indicate that despite fatigue he handled the situation very skilfully.

Another point that may be appropriate to the experiences gained by all three Comet crews on their ways to Australia is the enforced practice of flying a twin on one engine. One report on the race dismissed the difficulty by saying that a pilot had to travel a long distance with only half the intended horsepower available and therefore the aircraft must have lost a lot of speed. There is nothing untrue or otherwise incorrect in that statement, but it covers only a portion of the problem.

Firstly, unlike a single-engined aeroplane with an engine and propeller in the front, a twin has its sources of power on the wings outboard of the fuselage. When both engines are running at equal power the thrust is directly forward, as on the single, but if one engine is inoperative or working at reduced output the result is unbalanced. There is more thrust from the side of the good engine, which makes the aeroplane tend to turn (or more correctly, yaw) into the side of the dead unit. The increased speed of the wing on the outside of that yaw will generate more lift, so it will rise and cause a tendency to roll into the side of the inoperative engine.

When the airflow is passing the control surfaces at relatively high speeds there is no difficulty in keeping a straight heading on one engine with a light touch of rudder towards the 'live' side; in such circumstances many pilots have failed to detect power failures until the speed has decreased and the yawing effect becomes more noticeable. As the airspeed progressively lowers and the controls become less effective, more rudder is needed to counteract the power on the good side until it is possible to reach a speed at which *full* rudder is needed to keep straight. So far so good, just, but if the speed is allowed to decay further, only two possibilities are on offer: either to cross the controls by using the ailerons to keep the wings level in addition to using full rudder to maintain the heading, or to reduce power on the active engine and risk a resulting loss of height — or a bit of each. Although accepted practice on many of today's more docile aircraft, on earlier types neither action is a recipe for sustained flight, so care is needed early in the engine-out stage to establish a set of conditions that can be maintained for as long as is necessary either to reach the destination or, if available, an acceptable diversion. For this reason, height is a pilot's best friend, so when engine trouble developed on the crossing of the Timor Sea, Scott and Black climbed to 5000 feet from their earlier 1000 feet.

In addition to the problem previously explained about the Ratier propellers with their total absence of cockpit controls, they would have been unhelpful in the case of an engine failure. Today almost every aeroplane with more than one engine has feathering propellers, on which the blades can be turned beyond the normal coarse position used in cruising flight so that when the power unit is inoperative they are edge-on to the airflow and therefore generate very little wind resistance, or drag. If the Comet had a power failure, the propeller continued to act as a windmill and this would have an adverse effect on the aircraft's ability to maintain height 'on one'. A further factor that must have been nerve-wracking on long flights was the inability to stop an engine unless or until it seized. If oil pressure was lost or for any other reason the pilot elected to stop using a sick engine, because of the propeller's windmill action it would continue to revolve; this could aggravate the damage that had caused the first problem.

To add to these generalities about a twin-engined aeroplane of the Comet's concept, some of the few available words about the type itself may be of interest and these remarks were published in *Flight* dated 20th September 1934:

> The "Office"
>
> Accommodation is provided in the cockpit for a crew of two. They are seated one behind the other, and have dual controls so that they can take turns at piloting. A well-equipped instrument board is placed in front of the forward seat, and can be seen, by craning slightly, from the back seat also. The flying controls are of the usual type, with a plain "stick" for elevator and ailerons. A large wheel to the right of the front seat operates the undercarriage retracting gear. On the left is a lever which operates the air brake flaps fitted under the central trailing edge portion of the wing. Wheel brakes are connected to the rudder bar for steering on the ground. A transparent roof over the cockpit encloses the occupants.
>
> Flying control surfaces are of usual type, with a form of Frise aileron operated by the patented de Havilland differential method. The ailerons are mass-balanced by lead distributed along the leading edge of the aileron. Rudder and elevators have the usual bob-weight mass balances. In the rudder and elevator controls a mechanism has been inserted to give a very low gear ratio at small angles of movement of the control surfaces and an increasingly greater ratio for larger movements. This has, of course, been done in order to provide lightness of control and eliminate violent response to small movements of the controls at high speed.

Engine mountings and undercarriage attachments form almost the only metal parts in the "Comet". Welded steel tube construction is used for the engine supports, and the undercarriage wheels are carried on steel forks with telescopic limbs. The wheels are raised and lowered by a worm gear, or rather by the use of "overgrown" bolts and nuts, the "nuts" forming cable drums for the operation of the gear. Thus when the "nuts" are rotated by cables from the cockpit, they draw the "bolts" upwards, shorten one member, which forms one side of a triangle, and thereby raise the triangle and with it the wheel. As the worm gear is self-locking, no danger arises if the machine should land with the wheels not quite in the "fully down" position. The Dunlop wheels are provided with brakes, which can, as already mentioned, be operated separately via the foot bar. They can be locked "on" together for parking. Instead of a tail wheel there is the usual tail skid, which is thought to offer less air resistance and which helps to pull the machine up when landing.

The power plants fitted in the "Comet" are de Havilland "Gipsy Six" engines, specially modified for the purpose of the MacRobertson race. By using a modified cylinder head, valve rocker gear and piston, the overall height of the engine has been slightly reduced, the compression has been raised from 5.25 to 6.5 and the engine works satisfactorily on standard service fuel to D.T.D. 224 specification.'

Unfortunately no member of *Flight's* staff had been given the opportunity to fly this very special aeroplane and at that time no handling information was available for publication. The report continues with a brief explanation of the working of the Hamilton controllable-pitch propellers, then called airscrews in Britain, but eventually, following confusion in wartime with the word aircrew, the American term 'propeller' spread and this has become accepted universally. This information was prepared at the time of the first flights, for only G-ACSP flew briefly with Hamiltons; from test 8 on 26th September it was fitted with Ratiers and subsequently retained these. Both 'CSR and 'CSS flew from the beginning with the French units.
Nevertheless, although not really relevant until later in the Comet's career, it is of interest:

'In order to take advantage of the use of the Hamilton controllable pitch airscrews as fitted, the normal speed of the engine has been increased to 2,350 rpm. At the maximum of 2,400 rpm the engine develops 224 bhp on the bench. In the air this is slightly increased owing to the small degree of "boost" obtained by the high forward speed of the machine. A new crankshaft has had to be made to get an appropriate hub fixing for the Hamilton airscrew. The high pressure required (about 100 lb./sq.in.) to operate the pitch control of the propeller is obtained from the usual engine pumps through duplicate oil relief valves. It does not in any way affect the normal lubrication system. The standard arrangement of alternative hot or cold air supply for the carburetters has been retained in case adverse weather conditions are encountered during the race.

Warm air, if required, is taken from the vicinity of the cylinders through a flame trap.

A special coupling is provided on the rear end of the crankshaft for driving a rotary vacuum pump used to operate the Sperry gyro compass.'

To find much more detail about flying the Comet we need to turn again to the work of Flying Officer A. E. Clouston and to jump three years to September 1937. This information has been gleaned from a report that he wrote then, following his dramatic experience with severe icing over the Alps in the previous month. Although only extracts, these points give some insight into the pilot's view of G-ACSS's features and behaviour during that flight:

'Temperatures

These unfortunately are not known as no air temperature gauge was fitted to the Comet. The only temperature gauge installed in the aircraft was fitted to the oil system on the starboard engine and this dropped from 75° — 80° to about 60° — 65° in ice conditions at 19,000 feet. Both F/Lt. Nelson and myself had a fair amount of warm flying clothing, with silk-lined Service gloves, but our fingers and feet were numbed with cold. It is impossible to make any estimate of the air temperature during the flight other that to say it was bitterly cold and must have been well below freezing point throughout the cloud areas at 18,700 feet — 19,000 feet.

Ice

The ice that formed readily was of a glassy brittle appearance and hit the aeroplane in the form of a thick watery cumulus type of cloud. As the hail-stones or snow types of clouds were flown through, ice formed on the aircraft very slowly and with more of a powdery white appearance.

The ice was of the straightforward common types I have experienced frequently on the Northrop de-icing flights over England, and formed on every part of the aircraft presenting frontal area to the air flow.

De-icers

No de-icers were fitted to the Comet for de-icing the aerofoils or other surfaces.

Venturis

These were taken off the machine in view of my previous experience of their failures, and the port engine air pump was used to drive all the blind flying instruments.

Directional Sperry Gyro

This was driven off the port engine air pump fitted for that purpose. The instrument functioned satisfactorily except for a precession error of 10° — 12° to starboard in 15 minutes, which persisted through the flight. If the aircraft was banked more than 25° the numerical dial spun probably 80° — 150° before slowing up and settling down to record accurate turns. This Gyro was taken on loan from the R.A.E. and passed O.K. before fitting to the Comet.

Lateral Sperry Gyro

This was also run off the port engine air pump fitted for the purpose. It worked entirely satisfactorily and proved its worth. Had it not been for this instrument the flight would have ended in disaster. When the pitot head took on 1″ to 1½″ of ice and the A.S.I. ceased to indicate any mph, the lateral Gyro was used to keep the aircraft level fore and aft as well as laterally when blind flying through the ice areas. The Comet kept losing height slightly then regaining it by this method of blind flying.

Pitot Head

This carried about ¾″ of ice and functioned for over an hour satisfactorily in this condition. Ice eventually built up to 1″ — 1½″ when the A.S.I. fell to 40 mph for some minutes, then ceased to function altogether.

The Comet Pitot Head is of the small type and therefore more likely to be affected.

Compasses

The P.4 in the front cockpit functioned entirely satisfactorily and was successfully used on two occasions when the Directional Gyro began to spin out of control after a banked spiral (over 25°) when at periods the engines ceased to function temporarily.

The P.6 in the rear cockpit also functioned satisfactorily through the entire flight.

Engines

These are Gipsy VI Series II of 205 H.P. and 208 H.P. respectively. They are fitted out with independent flame traps or hot air controls to the carburettors. These controls were connected to separate levers in the pilot's cockpit. The engines were run at full throttle throughout the race with weakest maintained mixture. As well as using the altitude controls to advantage, overheating of the engines was checked by keeping a close watch on the oil temperatures as well as cylinder head temperatures, and using the mixture controls to advantage.

These engines ran perfectly for the entire flight until the hail and snow cloud belt was entered at 19,000 feet over the middle of the Swiss Alps.

The Comet had only entered these conditions some 4 — 6 minutes when both engines began to lose revs. slowly, from 2,100 revs. down to 1,000 revs. Assuming the carburettors were being frozen I immediately tried to bring in the hot air controls at the carburettors. These control levers would only move half their travel for some two minutes. By continually opening and shutting the hot air controls it was eventually possible to bring them fully on. The excessive stiffness and delay in getting on the hot air was caused probably by the leads being partly frozen in their guides. Once fully on I assumed the Gipsy VIs would immediately pick up to full revs. again, less about 80 — 100 rpm drop because the hot air control was in operation. This was not the case and although the altitude controls were closed and opened, likewise the throttles, and tried at all settings, the motors still remained at a tick-over. Height was lost to 13,000 feet where a break in the clouds afforded us clear air to spiral in and the engines almost immediately started to pick up slowly until full revs. were obtained and both motors ran smoothly, The Comet was obviously in a warmer belt of air although still bitterly cold, and the entire set of instruments fogged over inside the glasses. To read the blind flying instruments was then a concentrated job, and a little confusing.

Elevators

These operated normally throughout the flight and icing areas. They appeared to be unaffected by the formation of ice.

Rudder

This also worked normally throughout and the formation of ice did not change its efficiency.

Ailerons

During the climb from 7,000 feet to 18,000 feet through rain, hail and snow the ailerons became quite stiff to operate. At first I assumed this was my co-pilot just checking up on my blind flying, but as the control became stiffer until it was tiring to correct for the continual bumps we were flying through, I was satisfied after a discussion with F/Lt. Nelson that he was not touching the controls. It took me some time to prove this as the stiffness of the lateral movement was exactly the feel one experiences in a training machine when a nervous pupil tries to hang on to the controls and follow their movement, although told repeatedly by the Instructor not to touch them. I was so sure that George Nelson was trying to blind fly the Comet with me (he had no blind flying instruments in his cockpit) that I asked him to re-adjust the top radio set and by this observed both his hands free of the controls. The lateral control became stiffer and stiffer and I then considered it must be ice forming on the leading edge of the ailerons. On inspection of the ailerons before the flight I thought this impossible as the leading edge of the Comet aileron is so well protected behind the wing. If the ailerons were used viciously from side to side about ¾ of their travel (this was a very fatiguing procedure) they became almost freed up to their normal feel. This vicious lashing from side to side was done every 3 — 5 minutes while flying through snow, hail and rain at 18,700 feet. When my arms became tired I would hand over to F/Lt. Nelson momentarily to do the side to side lashing. By this means we prevented our lateral control becoming seized up altogether. At no time was the control seized up completely but wherever the stick was moved laterally it would remain jammed if left alone. When the engines failed and our second descent was made amongst the Alps, the ailerons became so stiff laterally that I feared jamming completely if this vicious lashing from side to side was continued to ease them up. I therefore left the lateral control in the central position. The control could now be described as almost seized up. By great force movement laterally could be made, but it felt as though it would jam completely and I considered it more dangerous to do this than to leave the control central and try and keep the Comet on an even keel by using rudder. After the descent into the Alps this control thawed out and although it became quite stiff later on flying through snow at 17,000 feet, the stiffness never repeated the almost seized stage again.

Stiffness of the lateral controls, leads and pulleys through cold could not produce such seizing as experienced on this flight. The fact the lateral control eased up each time we lashed the controls from side to side indicated ice on the leading edges of the ailerons being rubbed off, only to re-form again as soon as the normal aileron flying movement was reverted to. This formation of ice must be on the underside of the aileron as nothing was visible on the top side from the cockpit.

The magnitude of this experience is not exaggerated in any way, but just what actually took place is recorded here. I therefore consider its seriousness cannot be too greatly stressed. Aileron icing appears to be a serious hazard.

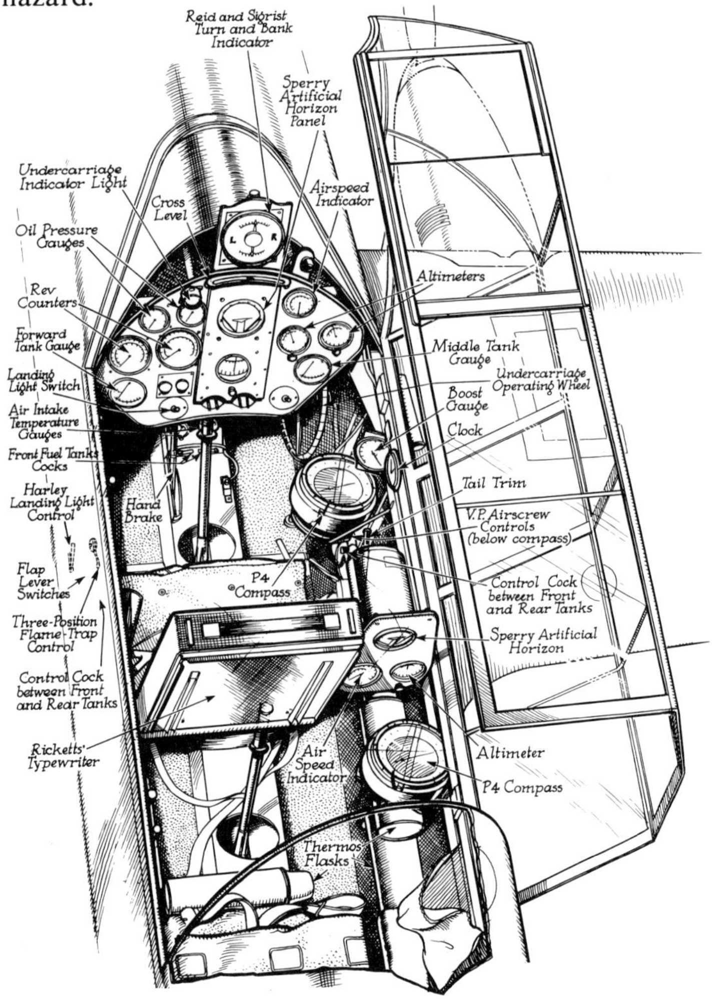

The cockpit of Comet G-ACSS when flown to New Zealand in March 1938 by Arthur Clouston and Victor Ricketts.

Airscrews

Slight vibration was evident from the engines climbing through the ice areas. As revs were constant and steady this was obviously ice forming on the propellers. Minor dull bangs were heard at intervals which indicated ice flying off the propellers and striking the fuselage.'

This covers the more technical aspects of the dramatic tale told in the previous chapter. I can think of no better way of ending a section on flying the Comet than these words from the person who may have known the DH 88 more fully and more intimately than any other pilot.

Chapter 8
Post-War Survival

By the end of World War II the sole surviving DH 88 was in a very poor condition. After it had stood outside for several years at Gravesend Airport the elements had won the battle of time and the old hulk remained untended to deteriorate still further. A few enthusiastic sporting pilots looked at G-ACSS in the hope of being able to restore the famous record-breaker to its rightful place — in the air — but each went away convinced that the task was too big to tackle.

Among possible purchasers who studied the prospect more fully than most was Geoffrey Alington, then a test pilot with the Fairey Aviation Company and a regular participant in national air races in his sleak single-seat Miles Sparrowhawk G-ADNL of 1935. To gain a more detailed technical opinion on 'CSS, he asked Ron Paine to carry out a thorough inspection. Paine was an exceptionally experienced engineer in the field of light and sporting aeroplanes and he features again later in the tale of the Comet, but at this stage let us say that his report was not very favourable. The aircraft *could* be restored, he said, but clearly the cost would be beyond the normal resources of an individual contender. So G-ACSS stood idle for a little longer, before being moved to Salisbury Hall, near St. Albans.

In 1950 the Festival of Britain was in the planning stages and items of national importance were being sought to display as examples of the best of the past from British industry. The Comet was selected, so work began on an external facelift. No-one had thoughts of preparing this famous historic aeroplane to fly again and presentation of a "pretty face" formed the basis of the plan. Unfortunately, though, through lack of space and to achieve maximum eye-catching impact, the aim was to hang the Comet from a roof, which created the need to remove every single component that would not be essential for the Festival display.

Not only were the engines removed, but almost every item imaginable was extracted in an intensive weight-reducing campaign. All the flying controls, fuel tanks, instruments, lines and leads disappeared and these were lost for all time; but that was not the only move that would be a handicap for anyone aiming to restore 'CSS to fly again: to provide the means for supporting the aircraft from the roof, holes were drilled through the wing spars!

In 1951 the DH 88 Comet G-ACSS – as restored for the Festival of Britain – stands proudly at Hatfield in front of a DH 106 Comet airliner.

I should add that the work in preparation for the Festival of Britain was planned with that alone in mind and clearly the removal of all unneeded items was an essential part of the task. Although commercial (and not aviation grade) materials were used, the restoration for the exhibition was carried out immaculately, largely under the direction of a de Havilland Technical School student, Tony Chalk, who unfortunately was killed in early 1966 flying as a test observer in a late production Vampire T.II destined for Austria. Most of the work on 'CSS was carried out in the Chester works at Broughton, after which a group of DH students from Panshanger went to the Festival site at Waterloo to assemble and tidy-up the aircraft for exhibition. However Tony Chalk's own description is the best way to relate the tale of the task. It was published in *Pylon,* which was the school's house journal:

'When the famous racing Comet G-ACSS recently appeared complete at Hatfield in its renovated form, many people were heard to comment on its beautiful lines and sleek appearance. Such remarks as "You can hardly believe it is that old," were quite common, but little could these people know of the tears (almost) and tribulations behind the renovating process, and it is this story that I am going to relate.

On my way up to Chester, by train, most of the near-by conversation was being carried out in Welsh, and I began to wonder what sort of a barbaric outpost I was being sent to, but on my arrival these fears were soon dispelled. Mr Cant welcomed me with open arms and began to look after me like a long-lost brother; he showed me round the factory, and then we had a look at the Comet. The first impression on seeing the relic was, "Ye Gods! What have I let myself in for?" It was indeed a sorry sight and, I thought, a crying shame that such a gallant aeroplane should have reached such a state of dilapidation.

It had been salvaged from Gravesend Aerodrome and before being taken to Salisbury Hall in 1944, it had apparently been lying in the open for about four years, during which time it had obviously proved to be an object of magnetic attraction to a large number of nasty-minded pigeons. Although the basic structure was in surprisingly good condition there was a great deal of external damage. The trailing edge of the wing was almost non-existent, with fabric and broken ribs dangling free and jagged holes here and there. The fuse' was lying on the ground with a crack in the keel through being dropped at some time, the nose under-panel was completely missing, so also was the most beautiful landing light, which I distinctly remember seeing at Salisbury Hall, and which presumably went as a souvenir like much of the aeroplane during its stay there. The canopy frame bore no resemblance to its original condition, being an ugly mass of twisted metal, but I was thankful to see that the windscreen was in one piece, this being of a most complex shape. The tail unit, although somewhat water-soaked on the underside, and presenting one or two small holes, would obviously not be much trouble. Unfortunately, this could not be said of all the sheet metal parts, engine cowlings, wing and tail root fairings, which were exceedingly mangled. The undercarriage assemblies were still complete and would be all right to refit after new tyres had been fitted.

After this preliminary inspection, I spent the rest of the day recovering from the shock, and next day set to and began the struggle. The wing seemed as though it would be the worst job, so I decided to get that off my chest first. With a gang of willing helpers from the Tech. School (anything is better than filing) removing the old dope and fabric, I began the laborious process of mending ribs and building up a new trailing edge section according to the best Panshanger tradition. The fact that the original drawings were almost illegible did not help in the least, and I soon discovered that it was quicker in most cases to work by eye. Regulations concerning repair schemes did not enter into it, so where a fabric patch would serve to prevent a hole from being noticed, a fabric patch was used.

The wing gradually began to take shape, and when it was complete the dope and fabric gang was called in to re-cover it; incidentally a few pounds of stopper had been applied to improve the surface. It is amazing what a multitude of sins can be covered by a little dope and fabric.

The fuselage was the next item on my agenda, and here again all the old dope and fabric had to be removed. Lightness was the watchword, as the machine was to be hung aloft for exhibition, so everything inside had to be stripped out — fuel tanks, controls, and even the original battery which was still intact. Holes were patched and cracks reinforced as on the wing, pounds of stopper and then fabric and dope were applied. With the aid of "a 'eavy 'ammer" the canopy frame was straightened, recovered with Cellon, and fitted.

Now, and my heart sank at the thought, the sheet metal items would have to be dealt with. The old paint about $\frac{1}{8}$ inch thick, had to be removed; thinners was found to be ineffective, but I eventually found some very potent paint stripper which did the job admirably, also removing most of the skin from my fingers. Much laborious tin bashing ensued, and eventually the parts were made to fit.

The undercarriage was fitted and wings and fuse, bolted together, and we began to see what the completed aeroplane would look like. All that remained to do now was the fitting of incomplete engines, an operation that was hampered somewhat by the lack of one engine bearer and some cowling supports which had to be made in the Tech. School. The engines had arrived some time before, and had already been stripped of all superfluous weight so an attempt was made to instal them — but no, the bearer feet were the wrong size! Frantic modifications, and the engines went in.

I should like here to place on record the invaluable assistance and advice accorded by representatives of the Engine and Propeller Companies. Anyone who has tried to make large-size engine cowlings behave reasonably will sympathize with my next task, they were "persuaded" to fit — 'nuff said.

Propellers were put on to dummy engine shafts, and that was that.

It had been arranged for the machine to have its final spray and polish at Hatfield, so all the good work had to come apart, and the pieces were packed onto two large conveyances. I was very thankful indeed when these arrived intact, as the tailplane had an overhang of two feet either side of the trailer. I should like to be able to write some of the driver's choicest comments concerning fog, brick lorries, lamp-posts, etc., but let it suffice to say that it was quite a trip.

With the aid of some of the "boys" from Panshanger the machine was soon re-assembled and ready for painting, but before this was done it was thought advisable to proof load the airframe. The wings and fuselage were loaded up with 2,000 lb. of shot bags, and the machine was slung from the roof, to see how it would like the idea of being suspended for about three months at the Festival of Britain Exhibition. I know at least one person who heaved a sigh of relief when there were no horrible creaks or groans. So off it went to the paintshop, and my work was completed.

In conclusion, I should like to take this opportunity to say how very much I have appreciated the assistance of all those who helped to give a tired aeroplane new life. In particular I am grateful to Mr. Cant for his untiring pains to make things run smoothly, and to his willing staff and protégés without whom I would have been lost.'

After the festival the Comet, which had been painted accurately in its original 1934 race-winning colour scheme and displaying its first name *Grosvenor House*, was dismantled and taken to Leavesden in Hertfordshire where it stood as the centrepiece in the de Havilland Engine Company's showroom.

G-ACSS lingered at Leavesden until 1965, when it was moved to Hatfield

'Grosvenor House' on display at Waterloo for the Festival of Britain in May 1951. So that it could be hung from the roof, holes were drilled in the mainspars.

After the Festival of Britain, G-ACSS was exhibited at Leavesden in the de Havilland Engine Company's showroom. One of the Gipsy Six Rs used in the MacRobertson race is exhibited beneath the Comet. This photograph was taken in 1954.

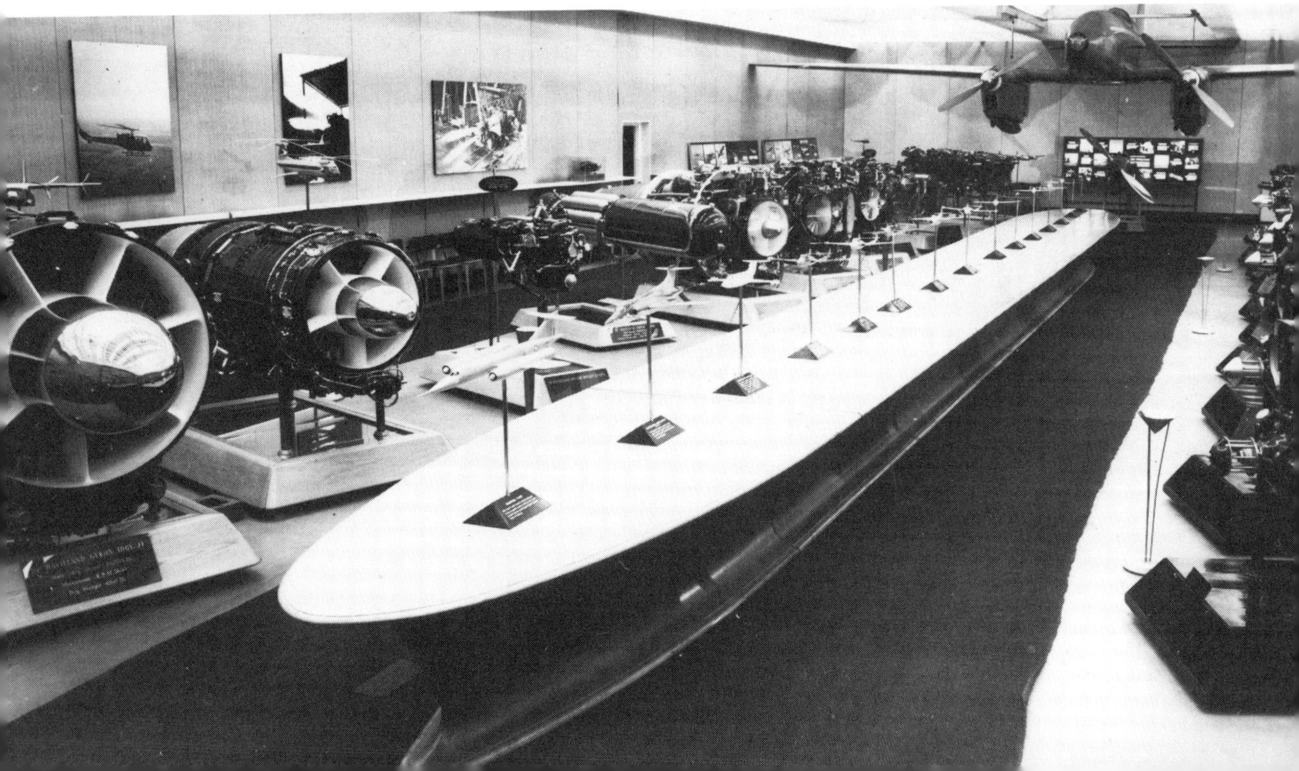

and some further face-lift work was carried out before the famous aeroplane appeared in the static park at the company's Open Day in July; there it stood in company with an example of its later namesake, the DH 106 Comet airliner. Afterwards, the DH 88 was handed-over to the Shuttleworth Collection at Old Warden Aerodrome near Biggleswade in Bedfordshire where it was delivered by road on 19th October 1965. A formal ceremony took place on 30th October, when *Grosvenor House* was received by Mrs. Dorothy Shuttleworth, the mother of Richard Shuttleworth, the Collection's founder who had been killed in an accident in 1940 whilst flying a Fairey Battle of the Royal Air Force.

Here perhaps I should mention that in 1928 Richard Shuttleworth had started gathering early motor cars and restoring these to full running order; four years later he extended his collecting activities to include historic aeroplanes and by the outbreak of World War II he had acquired an enviable selection of veteran machinery: among his items were a 1909 Bleriot XI of the type that had achieved the first aerial crossing of the English Channel, a Blackburn monoplane that had first flown in 1912 from the sands at Filey in Yorkshire (and which in later years became the oldest flyable British aeroplane in the world), a Sopwith Pup single-seat scout as used by the Royal Naval Air

On behalf of the Shuttleworth Collection at Old Warden in Bedfordshire, Mrs Dorothy Shuttleworth, mother of Richard Shuttleworth, receives G-ACSS from Air Commodore F. R. (Rod) Banks on behalf of the de Havilland Aircraft Company. This was in October 1965.

Service and the Royal Flying Corps from 1916 — and several others. Full details appear in a companion book *The Shuttleworth Collection.*

By 1965 when the Comet was added, the Collection at Old Warden had expanded to become the world's only range of flyable veteran and vintage aircraft with machines dating from the birth of flight to the second world war. Clearly, therefore, this was a most natural place for the Comet to spend its years of retirement.

The very smart, externally refurbished but internally incomplete G-ACSS proved to be a very popular exhibit at Old Warden. It shared hangar display space for all to see with other aeroplanes of historical note, such as the Percival Gull G-ADPR in which Jean Batten had made her record flights from England to Brazil, Australia and New Zealand in the mid-thirties. However, nearly all the Comet's stablemates were airworthy and when other aircraft were wheeled onto the small timeless grass aerodrome to take part in the Collection's flying day activities, often *Grosvenor House* was the only machine that failed to come out from the confines of its hangar. It emerged once or twice to be photographed and filmed statically, but its inert lifelessness showed as a contrast to the other machines which, occasionally and under strictly controlled conditions, would 'get up and go'. However, with no engines, no equipment and an airframe that had received no airworthiness checks since 1938, G-ACSS remained a static museum piece to remind visitors of great British achievements in 1934: achievements by the de Havilland enterprise which had designed, built and flown it within nine months from start to finish; and by the pilots who had battled their ways to Australia in record times and had overcome the many technical and weather problems that had faced them. *Grosvenor House* was very presentable externally and it became a regular talking-point among those visitors who knew of its past fame. At the same time other, often younger, people learnt about that fame.

In its static display state, *Grosvenor House* created the impression of being a more complete aeroplane than it was. Anyone who endeavoured to turn the propellers would realise that they were not even attached to crankshafts. They were nailed onto blocks of wood hidden behind the cowlings. A casual glance through the air intake for each engine would give no cause for suspicion, as the air cooling fins of a Gipsy cylinder could be seen just inside; but there was no more engine than that! The total absence of all internal fittings and fixtures was hidden from prowling eyes by a canvas sheet covering the cockpits. In short, what was not easily seen at a first glance was not there.

The presence of G-ACSS opened many people's minds to the whole Comet story. When did *Grosvenor House* fly last? What became of the French Comets? What happened to the Portuguese machine? Many such questions were fired frequently and only relatively recently have the fates of the foreign-registered machines become firmly established. The almost endless succession of queries raised by visitors to the Shuttleworth Collection proved the extent of the interest and that is how the idea started for the book that you are reading now.

As mentioned already in Chapter 6, the French-registered DH 88s were lost

'Grosvenor House' on one of its rare appearances on the airfield at Old Warden. Although smartly finished, G-ACSS was in bad structural and technical condition. This was in 1970, before the possibility of restoration had been considered.

in 1940 during a hangar fire at Istres and there have been no subsequent reports of any parts being salved, but as late as 1979 some parts of CS-AAJ (the former *Black Magic,* which had been flown in the Australia race by the Mollisons) were discovered in an outbuilding on a peasant's small-holding in Portugal. This was the first news since 1937 of the one-time G-ACSP and not surprisingly the remaining pieces were in poor shape.

Even if only a few small items are potentially usable either in themselves or as patterns, any parts of such a famous historic aeroplane have an instrinsic value historically if not in purely monetary terms. More surprisingly though, in 1935 not only had the Portuguese authorities obtained the aeroplane, but had acquired a set of drawings from the makers and most of these were still in existence! So, armed with some small pieces of fuselage, part of the tailplane, one seat, the battered remnants of sections of the wings and a set of photostat copies of many of the original drawings, an enthusiastic aircraft engineer named John Pierce decided to set-about the daunting task of rebuilding CS-AAJ/G-ACSP to flying condition.

Pierce was no newcomer to rebuilding ageing aeroplanes. Based on the Clwyd side of the English/Welsh border at Chirk, his Veteran and Vintage Aircraft Engineering Company had restored several types that had been out of production for many years and among the machines to emerge from his workshop were two smartly-finished de Havilland DH 89a Rapides. Work on the Comet seemed to proceed quite quickly and at one stage a photograph appeared showing the partially completed fuselage. In fact, this was the section that followed standard airframe constructional practice and none of the intricate cross-laminated sections had been tackled; clearly these would have followed at a slower pace. However, problems abounded. A fire in a barn adjacent to the hangar destroyed several items including the Comet tailplane and then the almost inevitable financial difficulties came to the fore. Pierce was unable to continue the Comet restoration work and in 1982 the project was abandoned, so now there is little likelihood that we will see a largely-new DH Comet in the air with the historic registration G-ACSP that the salved components would have entitled it to carry.

With the loss of the former *Black Magic* and with the unfortunate and known destruction of F-ANPY, F-ANPZ and G-ADEF, the world of historic aviation became aware that no new surprises could emerge. G-ACSS was the only DH 88 Comet in existence anywhere in the world and after so long a period of decay, with non-aviation materials used in the Festival facelift and with holes through the wing spars, this could not be described as a particularly healthy specimen; but 'CSS alone represented the tradition set by one of the most significant record-breaking aeroplanes that had ever penetrated international skies. Even in a collection of airworthy aeroplanes, a non-flying specimen can have a place of pride, especially if it is of such importance as the DH 88 and if there are good technical or other reasons why it should remain as a static exhibit. Such was the case with the Comet. G-ACSS, resplendent in red and white as *Grosvenor House*, was seen and admired as a museum piece even though its stable-partners were able to do what aeroplanes were designed to do: to fly.

Chapter 9
New Life for Old

"Well, why don't we?" This was one of those remarks that happened to be made one dark winter afternoon in one of the small hangars at Old Warden aerodrome. It could have been said at any time and it might have meant almost anything. However, the message went home and for a few minutes two people stood silently, each wondering who would speak or act next.

Even today, more than twelve years later, the scene is clear. The place was No. 3 hangar, in which civil and sporting aeroplanes of the years between the wars were housed and exhibited; the precise spot was at the back of the building, where G-ACSS stood. The only doubtful point is confirmation of who asked the question. Only two people were present, so there was an even chance of either party deserving the credit — or the blame. Wally Berry, the most experienced aircraft engineer at the Collection, was one; I was the other. We happened to be walking in opposite directions and for some reason each stopped, paused and looked at the Comet. The problem is that each thinks the other spoke first. By now the answer is irrelevant, for after we had recovered from the initial shock of the thought, we stood for a considerable time and just talked generally about the merits of the idea. We knew there would be plenty of problems, but for the moment these seemed to melt as we imagined the end result.

By now some of the technical problems surrounding the Comet will be well-known to readers, but the extent of the structural deterioration could not be determined until at least a provisional internal survey had been completed. In itself this created a new potential problem of an entirely different kind, for opening and stripping sections of the skin would entail total destruction of the smart paint scheme and if the results of the check proved unfavourable there would be just a tatty, partially dismembered Comet with a reduced likelihood of a sustained future.

Many discussions took place. Hawker Siddeley Aviation, as successors to the de Havilland Aircraft Company, could not muster much enthusiasm for the idea; many people at Hatfield were aware of the aeroplane's years spent rotting at Gravesend and, of course, they knew the nature of the work that had been carried out in preparing *Grosvenor House* for its sojourn hanging from the roof in the Festival of Britain. On the other hand, enthusiasm at Old

The Transport Trust was the first organisation to provide financial support for the restoration project. In June 1973 Ron Wilsdon, the Trust's Chairman (left), John Peyton, Minister for Transport Industries and Sir Peter Allen, the Trust's President, stand before G-ACSS at Old Warden when the Comet was being dismantled prior to the start of restoration.

Warden grew and preliminary inspections indicated that the task could be tackled. More through a combination of obstinacy and perhaps misjudged enthusiasm than through sheer management judgement, I approached the Transport Trust, as the organisation that had been established to help worthwhile selected preservation projects in any field of transport; the plea was succesful and the Trust's Council agreed to provide a base for the subsequent national appeal with an immediate cheque for £2000. Almost as this cheque was being signed, I was invited to lunch at a Hatfield hostelry where I was well fed, watered and advised of the folly of the proposals. This operation was carried out with carefully pre-planned precision and I found myself facing and hearing the concerted views of six selected people, including one who described in detail the non-aviation techniques that had been used in repairs to one wing-tip. I was quite unable to match the technical expertise that was lined-up for this indoctrination exercise, so I escaped from reality by claiming that I could do no more than report back to the Trustees. In practice, they had not been involved in any way at this stage, as I wished to present the project as a practical and manageable venture, which I was not then in a position to do.

In the search for funds I wrote and talked to many other people. Clearly though, care was needed in seeking money for a venture that was not yet approved and which still might prove to be impossible to achieve. However, by arranging a form of two-pronged approach, using the Transport Trust as the second prong, we managed a politically successful coup with Hawker Siddeley Aviation's head office in London. Sir Arnold Hall, the chairman, agreed to provide £5000 for the Comet restoration project; this was almost his final signature in his capacity as the head of this private-enterprise organisation before the HSA name disappeared into oblivion as a part of the new and nationalised British Aerospace.

Armed with the donation of £2000 and a promise of £5000, each from a well-respected source, I was in a stronger position to face other prospective donors, but this increased the overall responsibility over the need to be in a position to undertake the technical side of the commitment. Initially, this would have been hampered through total absence of funds; later, when the first cheques materialised, the position was reversed, for with a little money and several signs of more to come there could be no escape from the job of getting *Grosvenor House* back into the air.

By this time I obtained agreement from Air Commodore Allen Wheeler, then the Aviation Trustee of the Richard Ormonde Shuttleworth Remembrance Trust, which is the parent body of the Collection. He was surprised at the stage that the project had reached on an unofficial basis, but he accepted the situation as one with a future and very soon he approached his contacts in Australia to request more money. The arrangement within the Trust was that the Comet would be restored as far as possible on a self-financing basis; all possible help should be sought from organisations and individuals who would give money to the project and, when the engineering requirements were known in greater detail, appropriate companies would be approached to request their technical participation.

The Collection's determination to go ahead, with the substantial donation from the Hawker-Siddeley head office and signs of support from both industry and individuals, produced a very welcome change of heart at Hatfield. Once people knew that the project was to proceed, activity behind the scenes resembled that of a beaver colony.

A project of this nature calls for involvement by many people. As investigation continued, most people realised that the total task would be considerably greater than had been anticipated. Not long passed before the project became recognised as the most ambitious aircraft restoration task ever tackled by the preservation movement. Certainly, despite the many skills available within the small workforce at Old Warden, even if all the required funds could be made available, the Collection would be unable to manage the work alone.

An aeroplane may not fly unless it holds some form of authorisation to do so from the Civil Aviation Authority. Normally a machine has a Certificate of Airworthiness, which may be issued in the transport, aerial work, private or special category according to the nature of the flying to be carried out. However, subject to certain very sensible limitations, some aircraft may be granted exemptions from the requirements for full certificates and, instead, they operate on permits to fly. This does not mean that a machine is not fully airworthy, but enables certain dispensations to be granted, especially with an older type that had been designed at a time when some present-day requirements were not needed and which might be difficult or even impossible to incorporate retrospectively.

The CAA conducts this side of its overall activity through the Airworthiness Division, based at Redhill in Surrey. Although all major policy decisions must emerge from there, responsibility for detailed supervision of individual aircraft is delegated to a number of area offices. When the Comet work was beginning, Old Warden came in the parish of the Division's Hatfield office, although more recently this has been transferred to Luton. However, the same surveyor-in-charge has been involved throughout and the work would have been impossible without the whole-hearted co-operation of Colin Turner, who has been confronted not only with the problems directly relating to the extent and standard of the work to be carried out, but who faced also a change in the type of airworthiness category that would be required.

The early stages of restoration at Old Warden: the inverted wing of G-ACSS with all the lower planking removed so that the internal structure can be examined.

The wing with some of the new diagonal planking in place. Subsequently this needed to be removed.

Originally the aim was for the Comet to operate to normal Shuttleworth practices, for which a permit to fly would suffice, but then many external pressures mounted for the aircraft to fly to Australia to mark the fiftieth anniversary of the MacRobertson Air Race. Although at no time did this idea develop beyond the stage of being just a possibility, if it might fly through international territory on such a long route, a Certificate of Airworthiness would be essential and this called for a change of plan in the scale of the work that would be required. As an example, for flying only within the United Kingdom, a reduced fuel tankage would be adequate and in turn this would lower the maximum weight for which the Comet would need to be cleared. This would reduce the loads and stresses and produce a known improvement in the take-off, climb and single-engine performances. The projected flight to Australia, however, removed all these advantages, leading to a complete review of the airworthiness requirements to be laid down. Even some new design and stress calculations would be needed.

One of the first organisations to become actively involved on the technical side was Hants and Sussex Aviation of Portsmouth. The company offered to carry out complete overhauls of the engines if the Collection could find, obtain and despatch to their works sufficient parts from which two (or possibly three) power units could be assembled to rebuild standards. Suitable Gipsy Sixes were not available and a decision was reached to use the Gipsy Queen II, which was the Service version used in Percival Proctors. So a search began and appropriate engines were acquired from various sources: four (two of which had been used in one of the very few Rapide IVs) were presented by the Army Parachute Centre at Netheravon, one came from Derby through Ron Paine and another was bought for £150 from a Welsh coal miner who owned a battered Proctor that he had intended to restore but who had abandoned the idea. The six Queen IIs were given preliminary inspections at Old Warden to ensure availability of all components before they were despatched to Hants and Sussex for the major overhauls to begin.

Altogether nearly fifty organisations have provided technical help. Realising the historical significance and the national importance of the project, most have participated in the programme at no charge. Now, however, let us consider the details of the technical task facing all concerned on the practical side. For this, to ensure accuracy, I have sought and obtained considerable help from those who have been close to the project, including Ron Paine and Chris Morris.

After the decision to rebuild the aircraft to flying condition, in June 1973, an engineering report had been drawn up by the Collection's staff, detailing the areas of the structure to be inspected or rebuilt, the items missing and the problems likely to be encountered. Also this detailed the manpower required and the probable cost at the time of the external inspection. No time scale forecast was attempted and, looking back, this was very sensible for such a large, complex and unknown project.

On the right is a new section of fuselage skinning showing how the double curvature has been achieved by diagonal planking. In the centre is the jig.

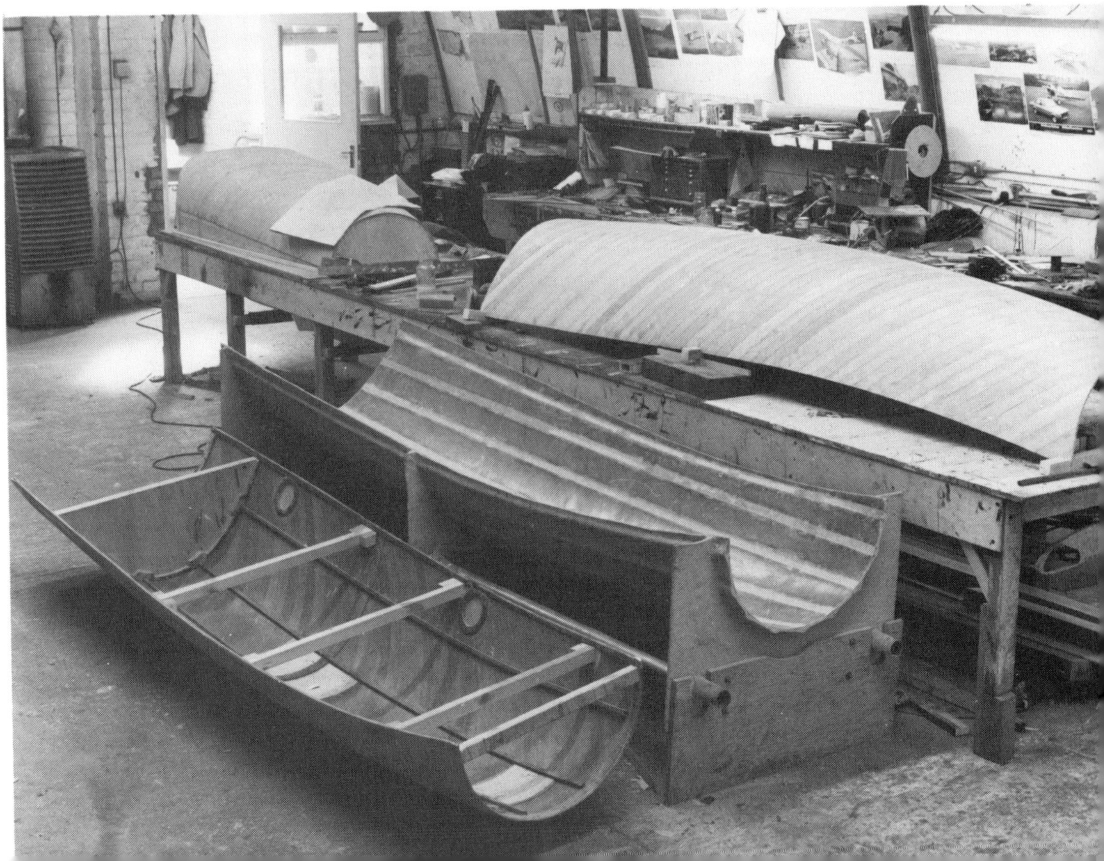

With the aircraft externally complete and on static display within the Collection, energies were directed from early 1974 in obtaining all available drawings via Vernon Clarkson, Product Support Manager at Hawker Siddeley, Chester, but some had been lost in war-time bombing and, over the years, many others had been lost or mislaid before the remainder were put on microfilm. Several important general arrangement plans were not available, leading to a great deal of extra work. Les Geering, Chief Stressman of Hawker Siddeley Aviation at Hatfield, was willing to help with the project, by re-stressing any alterations required and bringing the original calculations up to date. Fortunately most of the calculations were still available in the Hatfield archives. Other design specialists at Hatfield were contacted and all were keen to help. The undercarriage oleo legs were sent to Dowty Rotol, of Cheltenham, for a complete strip and rebuild. Dunlop were contacted regarding the tyres, tubes and oil hoses.

At the end of the display season in 1974, the aircraft was dismantled and the one-piece wing of 44 feet span was brought into the workshop, where on numerous trestles it was carefully supported in the inverted position. To achieve this, as the working area was only forty feet long, the engineers offered to forfeit their rest room at the end of the shop; the door was removed, a section of wall was opened and the wing-tip of 'CSS replaced the table and the coffee cups.

The double diagonal planking was difficult to remove in places for, like the curate's egg, it was good in parts even though the old, moisture-prone casein glue had been used throughout. For a thorough look into the enclosed structure, some of the undersurface and leading edge ply also was removed.

Various rib-to-spar brackets, undercarriage brackets and fittings to the fuselage were checked, repainted and reassembled with some new bolts, while

Some of the many components that have been manufactured to replace unserviceable or, in many cases, missing, items.

Some of the new fixtures and fittings installed in the cockpit. The large wheel on the right wall is for raising and lowering the undercarriage.

the trailing edges were repaired and some ribs aft of the rear spar, that were fabric covered, were renewed as required. A ply panel on the front face of the front spar in the centre section was removed and the spar structure deemed satisfactory, although the spar booms were spliced in this area with the same 40-years-old glue. The holes drilled for the Festival of Britain slinging points were repaired to a specially approved scheme and various ribs were modified for the new type of retraction system. The inside of the top surface planking looked reasonable, so after a protective coat of red dope, the long task of laying the varying width and thickness spruce laths or strips over the prepared surface began. Before this stage was reached a great deal of spruce had been obtained from two sources, including one large baulk that was airlifted from Canada at no charge. This was converted to aircraft use, which meant inspecting and cutting the baulks to obtain the correct grain direction, with the minimum amount of waste. The moisture content needed to be within limits and the wood was checked to ensure that there were no defects.

After the wing had been finally checked for truth on the trestles, the planking began in early 1977. After the first layer had been applied with Aerodux, made and supplied free of charge by Ciba-Geigy of Duxford, the gradual steps in the planking were sanded to a smooth contour, ready for the second layer the other way, again sanding down by hand to give the clean aerofoil section. Several coats of dope and a layer of Irish linen to seal the structure, then more coats of dope, left the surface glass smooth. One leading edge was left off at this stage to allow the airspeed indicator pipe-work to be installed at a later date.

During the wing programme, the fuselage also was being worked on, to replace decaying woodwork at the bottom of the structural bulkheads, where moisture had delaminated the joints. As all the control gear had been removed in 1951 and no complete drawing set existed, the control box on the floor between the pilots was partly redesigned and completely made from scratch. Dimensions were taken from the fuselage, the part designed, made and fitted; this is called 'reverse engineering'. The design engineer from Hatfield who took on this task initially, out of sheer enthusiasm, was H.M. 'Bob' Roberts, of the Mechanical Design Section. Throughout the project the majority of his work has been carried out in the evenings and at weekends. To enable him to continue at home, the Collection built him a portable drawing board to his design.

The flap lever and quadrant mechanism, elevator trim bias, manual override for the undercarriage retracting mechanism, the instrument panel attachments, seats and attachments, a complete new cockpit canopy to replace the lightweight one fitted for the Festival and hand brake lever system were all to be designed, drawn or at the least re-drawn, before they could be made. Several items were changed time after time, because of conflicting routing paths or other requirements. Later, the instrument panel and supporting structure was completely redesigned and remade, taking almost a year to get right. For modern requirements extra items had to be added in an already crowded cockpit, such as pitch controls for the propellers, mixture controls for

The original undercarriage retraction system relied on the cable revolving evenly round the drum. This shows how easily the results of counting the turns on the handwheel could be incorrect.

the engines, a master switch for the battery, a fire extinguisher, and the radio aids, with all their accompanying control boxes and indicators. Some items had been fitted in the past, but so far no drawing or photo of the port side of the cockpit has been found.

The ailerons, tailplane, elevators and rudder were all ply covered, with no inspection patches for internal checks, so one or both of the ply faces were systematically removed, the internal structure repaired or reprotected and new ply fitted. Attachment brackets and bolts were replaced and checked as required. The ailerons and the flaps also were inspected and one or two of these needed a complete rebuild. The aileron gear was remade at Hatfield, because of cracks found in the original fittings. At this stage Hatfield's apprentices were making or machining all the new components that Bob Roberts was either drawing or redesigning, with Les Geering clearing the stress side. The fuel system was redesigned by Eddie Livermore, the Chief Power Plant

Wally Berry, the Collection's most experienced aircraft engineer, prepares a Gipsy Queen II for trial installation. The engines had been overhauled free of charge by Hants and Sussex Aviation.

Systems Engineer, but utilising the same system of three tanks with a slightly reduced centre tank capacity to make way for the retraction motor. Later, for various reasons, the rear tank behind the cockpit was deleted, but this was mainly to make way for the radio gear. The undercarriage frames had endured hard use and abuse in the past, so it was with no surprise that the apprentices at British Airways (Overseas Division) under the direction of R. G. Mitchell, the Engineering Superintendent at Heathrow, needed to scrap three of the four and remake them from some old and some new drawings; this followed a redesign in conjunction with Hatfield to increase the strength of the structure and to accommodate the new operating mechanism. Subsequently, the attachment and alignment of this structure to the wing fittings proved quite a challenge to several people.

On the fuselage, some of the skin beneath the forward section appeared to be suspect and this was peeled away for more detailed inspection. Here the damp had penetrated the original casein glue attaching the plywood skinning on the

Ron Paine, the project co-ordinator, seems pleased with the progress in the cockpit while Bob Roberts, of British Aerospace, views the results from a distance.

The Comet at an advanced stage at Old Warden. Subsequently it was dismantled for further wing repair work and for despatch to the Royal Aircraft Establishment at Farnborough.

longerons, but below this the adhesion had remained so secure that the skin would part from the longerons and stringers only by planing! By now, as the total task was developing into a complex operation far beyond the originally conceived programme, experienced overall supervision was needed. The small engineering team had about twenty-five historic aeroplanes to maintain and whenever an unexpected snag arose someone with the necessary skills would be diverted to tackle the task. This tended to cause a loss of both concentration and continuity on the Comet; although unacceptable, it was understandable, for with an aeroplane on a restoration programme that was spread over several years the feeling of urgency was at a low ebb. This would compare unfavourably with the need for instant action on an airworthy aeroplane that might be destined to earn its keep at a flying display and which needed rapid rectification of a short-term fault.

The result was the appointment of a Comet project co-ordinator. An obvious candidate was Ron Paine, who had learnt his basic skills as an apprentice at Brooklands in the late twenties and who has devoted his entire working life to the practical sides of aviation. A licensed engineer and a pilot, he had been Technical Director of Wolverhampton Aviation and Derby Aviation and, later, Managing Director and Chairman of British Midland Airways. Throughout his time as an executive, he had continued to get his hands dirty and at one time he set-up the small-scale production of the all-wooden Gipsy Gemini from incomplete airframes that had been left over when Miles Aircraft had fallen into liquidation. Also, he had maintained and continually improved the performance of his own Miles Hawk Speed-Six G-ADGP, which he raced successfully in national events for many years.

Fortunately Ron Paine was semi-retired and able to devote both time and energy to the cause of the Comet. With no other aircraft on the Shuttleworth fleet to consider, he brought a new and disciplined approach to the task and he called for much more detailed investigation into every item of the structure. Some people felt that he delved too deeply and removed sections that seemed to be sound, but he discovered a number of hidden weaknesses and his safety-first philosophy proved to be right.

Although more glue joints and the wood around them revealed their age, much of the structure's survival and reluctance to come apart could be traced to the wood screws and gimp pins that Geoffrey de Havilland had incorporated on all Stag Lane products. Screws every four inches and pins at intervals of only one and a quarter inches had strengthened not only the Comet but also all the Moths; this practice was followed later by several other manufacturers.

In the main, traditional materials have been used in the restoration work, but where modern alternatives offer better prospects for long and reliable service, minor concessions have been allowed. Instead of using the casein glue of the thirties, the joints are being held together by Ciba-Geigy's Aerodux, while the original and badly corroded nose cone made in Electron magnesium alloy has been replaced by a glassfibre specimen produced by Specialised Mouldings of Huntingdon. The same company has produced the pilots' seats,

One stage in the development of the instrument layout on the cramped panel. The planning of the cockpit, to include space for the modern radio equipment, has been a major task.

using the surviving original from G-ACSP as a pattern. Another break from the original specification is the use of a tailwheel instead of a skid. This is essential for operation from modern airfields with hard runways and certainly will give the aircraft and its occupants a smoother ground ride.

Perhaps the most noteworthy changes from the original specification concern the operating side. Already I have mentioned some of the problems and the accidents relating to the original hand-wound undercarriage retraction and lowering system, so all concerned agreed that the risk of failure was too great to allow total reliance on this. A working mock-up of the early arrangement has proved useful in the research needed to overcome this problem, but the final result is the use of an electric motor (believed to have come from a Buccaneer canopy) and a flap torque tube from a British

The new fuel tanks built by British Aerospace alongside the partially completed airframe.

Aerospace B.Ae 146! Part of the old system will remain operable, however, using the large handwheel in the front cockpit and a manual override; this will serve as a standby in case of any electrical problem, but the old continuous cable and drum that gave all the early problems have been deleted.

Another change from the original equipment of 1934 is the use of de Havilland controllable-pitch propellers, but these are identical with the units fitted to 'CSS in 1937. The original Ratiers were not available and, as with the undercarriage system, their reliability left much to the imagination. The propellers to be fitted were removed from Percival Proctors (which used Queen IIs) and at the start of the project three of these were cropped and completely overhauled to nil-hour standard by Hawker Siddeley Dynamics at Hatfield. A fourth, as yet uncropped, propeller awaits conversion with Hants and Sussex.

Late in 1979 the fuselage was transported from Old Warden to Hatfield to have several systems tackled nearer to the pulse of the project. The newly-made fuel tanks were offered to the fuselage, with a trial installation of the vent pipes, filler necks and tank slings for measuring purposes. Another major job was to adjust and fit the newly-manufactured canopy, with panels supplied by Triplex Safety Glass. One of the Collection's engineers went to Hatfield to help with the fitting of the various components.

This was the stage at which some people thought it feasible that if the project could be completed by 1984, the aircraft had a chance to re-enact the Melbourne Race of fifty years previously; however this time it would go on many easy and shorter stage lengths, following the original route as closely as possible and in company with other aircraft. To this end, an electrical and radio system was designed, with a generator driven by a flexible drive from each engine, a lead acid battery in the nose compartment and VHF, ADF and

VOR/ILS. This would enable the aircraft to fly in good visual daytime conditions, with a limited bad weather capability. To this end, various appropriate firms were contacted and Marconi of Rochester were willing to supply and fit all the radio equipment, the wiring harnesses, control gear, all the hidden aerials and most of the mounting trays; in many cases they redesigned and altered their own equipment to suit the wooden structure of the aircraft as all modern sets, of course, are fitted in entirely metal surroundings.

In 1981 when the wing was structurally completed, with the new access holes for the undercarriage motor and torque tube access reinforced, the wing was moved from the woodwork shop into No. 6 hangar where a corner was sectioned off with a polythene and Dexion structure. Although time-consuming, this construction was necessary because of the likelihood of having to work on the airframe during winter, requiring heating. Also, it would keep the area free from dust and birds, which would be a necessity for spraying. The wing and fuselage were supported by jacks supplied by Mingay's of Chigwell and a scaffolding walkway was built to a height that would give easy access for the engineers and enable them to carry out trial undercarriage retractions. All systems started to gel at this stage: the undercarriage and engine bearers were temporarily fitted, whilst special bolts and spacers were being made; also slave engines were installed, to design and measure controls, fuel line runs, oil pipes, electrical routing, positioning of tachometer generators, starter solenoids, vacuum regulators, voltage regulator control boxes and suppressors. This avoided the need to use the Gipsy Queen II engines that had been completely overhauled free of charge by Hants and Sussex Aviation of Portsmouth, and held by them in ideal storage conditions at the Collection's request.

The flaps were put in place to measure and manufacture the attaching linkages and the tail unit was assembled, rigged and connected to finalise the routing of the cables. The instrument panel (fifth design?) had been made at Hatfield by a Shuttleworth apprentice on loan, the same apprentice later making both the main radio consoles mounted just forward of the front control column and the cable brake system with its rudder pedal bias. The final routing was resolved very late in the project.

With the undercarriage structure installed, the oleo units and retraction rods could be offered up and altered as required, to ensure efficient and friction-free movement. Then the right angle drive box was connected to the torque tubes and the electric motor, but as no electrics were available at the time, the hand-wheel was used for the initial trials. Wheels, tyres and brakes that had been overhauled or made new by Dunlop, were attached, with new brake cables manufactured by Bowden. Hand retraction tests inevitably showed up tight and binding areas and this called for many trial adjustments of the tube frames.

Away at the time for refurbishing were all the wing root and tailplane fairings, the engine cowlings, and the retracting mud flaps/undercarriage doors. In 1934 they had been made from aluminium magnesium sheet, prone to cracking, and had myriads of screw holes round most of the edges.

In the background at Old Warden much work was being done in obtaining missing items. Requests on a large board near the airframe in No. 6 hangar often brought wanted items: spare propeller blades, many Proctor, Rapide and Chipmunk items on the electrical side, such as flexible generator drives, generators and control panels; also oil temperature gauges, induction temperature gauges, altimeters, navigation lights, oil hoses and breather pipes. Spare plugs also were a requirement, as were tailwheel tyres and tubes. Each item donated warranted a letter of thanks, and each item crossed off the list was replaced by another still required for the project. On the face of it, one of the hardest items to locate should have been the easiest: the landing light bulb. Several have been seen in Viscounts, but understandably the owners were reluctant to part with them.

All was going well, albeit slowly. Several projected target dates had been broken; in 1979, 1981 had seemed a possibility and at the start of 1982 completion seemed feasible by Spring of the following year. Under the part-time but essential supervision of Ron Paine as project co-ordinator, two apprentices and two woodwork personnel plus a licensed engineer were working on the task for a substantial portion of the time; more sections of industry were really beginning to help in kind (see appendix), most parts were materialising and to the casual bystander the whole project started to show positive progress.

Then major problems began to appear. In 1981, close inspection after cutting back some of the fabric already fitted revealed some tell-tale water stains, which due to lack of experienced supervision, had not been heeded in the early stages; the original top planking had shown some minor movement on the outside laths near the ailerons, but in Spring 1982, poor adhesion was found in the leading edge area of plywood that had not been renewed at the beginning of the programme. Further inspection in these areas revealed more separation of the glue faces, where water had collected in the leading edge during storage in the open. Delamination was found between the top skin planking and the front spar in small areas, so it was decided to remove the fabric covering over the planking. Further water soakage was apparent and so in October 1982, the wing was split from the fuselage (a job that all concerned hoped might not need to happen again during the rebuild,) and transported back to the woodwork shop. All the top skinning was removed, showing that the decision was correct and bringing to light even more areas of degeneration of the old glue. Ply faces from the box spars were removed (showing a small amount of water soakage in local areas) and relieved clamping pressure, showing up any joints in the spruce spar booms themselves as the primary structure. Several repair schemes were raised by Hatfield and the smaller ones were carried out. Then, on Thursday 11th November 1982, the Trustees of the Collection had no alternative but to put the whole project temporarily 'on ice'. This was because of the structural problems already mentioned and also because of the extra cost that would be involved. Although technical help was available from more sections of industry that anyone had dared to anticipate at the start of the project, much-needed financial support from well-wishers had fallen far short of expectations. As a result, Old Warden was needing to fund the project from normal running income, but this was contrary to the terms of the original agreement by which the Collection would restore the Comet for the nation. Also, for the long-term future this would be self-defeating and could not be allowed to continue. Personnel were withdrawn from the project, but one engineer carried on, hoping to stabilise the wing for storage over an unknown period.

Several organisations were approached and after initial negotiations by Dr. Alan Roscoe of Shuttleworth's Executive Committee, the Royal Aircraft Establishment at Farnborough agreed to take on the project, mainly as an exercise for craft apprentices. In May 1983 the whole task, fortunately with the goodwill of the supporting firms, moved to the RAE.

The move to Farnborough entailed a change of CAA "parish" from Luton to London Airport (Heathrow), where the project came under the surveillance of Cyril Cresswell, the Senior Surveyor at that office. The entire past history of the aeroplane and its records were studied and one of the first moves considered to be necessary was to open the front spar web and to replace or rework all suspect timber. The port wingtip, the earlier damage to which was mentioned in Chapter 6, was re-examined and given further extensive repairs. The upper and lower surfaces of the wing were replaced and many new metal fittings were produced. Whilst this work was being carried out, to ensure that the wing retained its correct shape, it was placed in a jig, for which surface plates were used. The workmanship at Farnborough has proved to be second to none and the people engaged in the task have created an atmosphere of both professionalism and enthusiasm combined.

Throughout this period the Shuttleworth Collection has paid for all items used; Ron Paine has continued to provide specialist technical knowledge and the liaison between the RAE and the Collection has been maintained through Wing Commander R. F. (Dicky) Martin, who is Old Warden's most experienced pilot and a member of the Executive Committee with a responsibility for engineering. High level contact between the many parties involved has been established and maintained by Sir John Charnley, formerly the Controller of Research and Development at the Ministry of Defence and now a very active member of the Shuttleworth Collection's Aerodrome Committee.

It must be clear to all that the entire project has been bursting with technical problems. Extensive and unequalled experience with other old aircraft tackled by the Shuttleworth Collection over more than fifty years has shown how much more difficult is a restoration task than the construction from scratch of a new, replica or reproduction aircraft. There are many conflicting criteria: what to save, what to repair or rebuild and what to scrap. The left hand of historical authenticity says that the aim must be to retain as many original components as possible, yet the right hand must think in terms of airworthiness, safety and longevity. Probably there are few if any drawings and invariably the missing sheets are those that are most needed. No parts are readily available. The airframe and all its components may have doubtful backgrounds and the only way to carry out a thorough check on health is to open everything; yet once a structure is fully dismembered there is little likelihood of many items being re-usable, so then new parts are used and the originality is lost. So it revolves in a vicious circuit. These difficulties arise with every major restoration project, but in the case of the Comet they have reached unprecedented peaks.

Although the technical difficulties have been immense, I would be wrong to overlook other problems that have arisen. This is intended to be the true tale of the Comet and not a book about the pettiness of personal politics, but such a large-scale venture has involved not only dozens of organisations but considerable numbers of individuals: some say too many. Therefore to complete the portrait I must mention briefly the human hazards that have emerged.

Unfortunately some sections of the aviation press have not helped. On the assumption that fire and smoke are inter-related, no-one can deny that some mistakes have been made. The original inspection process in the first stages of the restoration programme should have been more comprehensive, for subsequent checks have revealed the structure to be less sound than was thought originally, but some reports about this came from marginal sources and the situation became exaggerated.

At one stage an announcement that G-ACSS would re-enact the 1934 air race by flying to Australia appeared; within a week I had received about twenty telephone calls and a dozen letters of objection from well-wishers who were concerned for the aircraft's long term future. Yet at no time had anyone, anywhere decided that the aircraft *would* fly half way round the world; in response to requests from numerous organisations and individuals in Australia, the Shuttleworth Collection's Aerodrome Committee undertook a feasibility study and the verdict was that it would not go — unless it did so aboard a larger aeroplane.

To reach this decision, a thorough investigation had been carried out into both the practicality of the plan and the technical specification to which the Comet should be completed. The latter included installing two 500-watt generators, not only to cope with the electrical load for the avionics, but also to ensure success for the electric starters (not previously fitted) which were considered necessary because of the difficulty in finding people along the route who would be qualified for hand swinging propellers. An electrically heated pitot head was another measure added for sense and safety. The overall route plan was worked round stage lengths of less than 1500 miles, a take-off weight of 5,500 lbs and average daily flying not exceeding five hours. All this involved some detailed research under the overall direction of Marshal of the Royal Air Force Sir John Grandy, who enlisted among others the well-known former test pilot Wing Commander R. (Bee) Beamont and the technical expertise of Ron Paine and Bob Roberts.

As a result of this, in February 1985 the partly restored but structurally complete aircraft was loaded into a Boeing 747 and transported to Melbourne, Australia, where it was on static exhibition at a number of locations as part of

The DH 91 Albatross followed the Comet in its clean lines and in its wing construction.

the celebrations for the 150th anniversary of the State of Victoria. This operation was made possible through the extensive efforts by RAE Farnborough; throughout its time away from home the Comet was under the active direction of Ron Paine, with Alan Hancock's assistance.

Three months later, the aircraft was returned to the UK, the wing and tail surfaces going to Farnborough and the fuselage to British Aerospace at Hatfield, for completion of all outstanding work and eventual despatch of the Farnborough components to Hatfield for final assembly, installation of systems and rigging checks.

By now a volunteer team of enthusiastic retired BAe personnel had been organised and among them they had the skills and expertise necessary to complete the aircraft to full flying standard. The team was supplemented by specialists from the various manufacturers and also had available the manufacturing resources of BAe Hatfield. Throughout this work The Shuttleworth Collection, as the aircraft's owners, provided extensive technical back-up.

Work continued throughout 1985, 1986 and into 1987 and gradually G-ACSS began to look like an aeroplane that would really fly again.

As the great day drew closer, enthusiasm increased enormously and on the wet and windy morning of Saturday 28th March 1987 the aircraft emerged from the hangar. The Gipsy Queen II engines with their DH constant speed propellers were run for the first time and the opportunity taken to check the functioning of the new comprehensive electrical system, the modified fuel system and the gyro instrument installation. Also, a short taxying test was carried out.

As was to be expected, a few problems were identified, but these were to be rectified during the following month while the last details, weighing and final inspection were completed.

Saturday 16th May 1987 was a very busy day, with the prospect of flight now clearly in view. Final inspections were carried out, engine runs and system checks including fuel flow checks were complete and final documentation prepared for submission to the CAA so that a Test Permit could be issued. By early evening the aircraft was presented to George Ellis, BAe Test Pilot and regular Shuttleworth pilot (who had the advantage of being in current practice on the UK's only airworthy Mosquito), to enable him to carry out his own engine runs and system checks and finally taxying tests before the aircraft was returned to the hangar for the night.

Sunday 17th May dawned bright and breezy. Fingers were crossed for the possibility of a first flight. After final adjustments to fuel load and ballast, George Ellis was aboard again, this time with parachute. Weather was not ideal, as clouds were building and the wind increasing, but after two fast taxying runs along the runway G-ACSS lined up for the third time and soon was airborne again under its own power and supporting its own weight for the first time since the last recorded flight on 4th July 1938.

This was an emotional and exciting time for all concerned, none more than for Ron Paine, largely through whose dedication and determination this restoration of a famous aircraft to full flying condition had been made possible.

This seems an appropriate place in which to give a brief, but up-to-date, assessment of the Comet as a flying machine and these are the written words of George Ellis which he prepared shortly after the first few test flights:

> 'Taxying on a hard surface is very difficult. The view is very restricted and the pilot cannot see within about 13° of straight ahead. Steering is by differential brakes, set by a hand lever and operated through the rudder pedals. Take-off is tricky, mainly because of the restricted view. The pilot must work hard to keep the aircraft on a straight line, but the restricted view makes it difficult to see any small change of heading.
>
> In the air, the aircraft has a mixture of good and bad features. On the credit side, directional stability is good, which was a very pleasant surprise, and this must have been an important factor in making it

The DH 98 Mosquito owed much to both the Comet and the Albatross. This was the prototype of the fighter version and the third Mosquito to be built.

bearable for pilots to manage on very long flights. The cockpit is relatively quiet and draught-free. The less pleasant features are that longitudinal stability is poor, the secondary controls are difficult to reach, and the view is poor: even in cruising flight one cannot see directly ahead of the aircraft. The Comet is very clean and picks up speed rapidly. Cruise speeds in excess of 180 mph are easily attained.

Landing is made difficult by several factors. The first is the poor view, which means that a curved approach must be flown. The second is the poor low speed characteristics (wing drop at the stall and sensitive elevators) which mean that a wheeler landing must be made. The aircraft is very clean, and is reluctant to slow down, so the pilot has a lot of work to do to keep straight as speed gradually reduces.'

Preliminary flight testing continued satisfactorily from Hatfield and the CAA granted special dispensation for G-ACSS to make its first post-war public appearance at the Mildenhall display in May. Wind and weather prevented the first two planned appearances, but on Sunday 24th the bright red machine returned to the place from which it had taken off at the start of the MacRobertson Air Race to Australia nearly 53 years previously. A week later Grosvenor House took part in a display at The Shuttleworth Collection's own all-grass aerodrome at Old Warden.

Clearly an aeroplane of this nature might suffer from some second-generation teething troubles. The most significant problem was a tendency for both engines to run at uncomfortably high temperatures, even at normal cruise power settings, but research and experiment led to a redesign of the air intakes and initial tests revealed a welcome lowering of the operating figures. Another difficulty was a slight tendency for the propellers to over-speed, leading to a need to adjust the range of pitch settings. However, on Saturday 4th July, following a display at Hatfield Open Day, G-ACSS was landed in almost windless conditions, which can be very tricky with swing-prone aircraft, and at about 30 mph a swing started and this developed into a ground loop, causing annoying — but repairable — damage.

Throughout its pre-war flying career the DH88 revealed some tricky characteristics and these came to the fore in several landing accidents, but the problem today is compounded by the need to operate from hard runways and the related reduction in the availability of directions in which to take off or land. The tailwheel that replaces the original skid increases the tendency to swing and the freedom for this to caster aggravates the situation. There is no recorded evidence of any aeroplane with a wooden fuselage being fitted with a lockable tail-wheel fork, but as this goes to press some final stress calculations are being prepared to determine whether this will be a practical possibility. This is just one of the many problems encountered in this exacting operation.

The famous DH88 Comet G-ACSS has entered a new phase in its flying life. It has appeared at three displays and is out of action only temporarily. Shortly it will be seen again where it belongs — in the air.

Appendix 1

MACROBERTSON AIR RACE

England to Australia: October 1934

LIST OF COMPETITORS

S = Speed H = Handicap

Racing No.	*Nominator*	*Aircraft and Engine*	*Pilots*	*Races*
2.	**Manawatu Aero Club** New Zealand	Miles Hawk Gipsy Major	Squadron Leader Malcolm MacGregor Henry Walker	H.
3.	**Lord Nuffield** Nuffield Place, Huntercombe, Henley-on-Thames	Airspeed 'Envoy' A.S.6 Wolseley A.R.9	G. E. Lowdell Flight-Lieut. D. F. Anderson	H.
4.	**Harold Gatty** 4416 Klingle Street, N.W., Washington, D.C., U.S.A.	Douglas Transport DC Curtiss Wright 'Cyclone'	J. Frye H. Hull	S. & H.
5.	**Col. Roscoe Turner** Whittier Hotel, Detroit, Mich.	Boeing Transport Pratt & Whitney 'Wasp'	Col. Roscoe Turner Clyde Pangborn	S.

Racing No. 5 – Boeing 247D.

Racing No.	Nominator	Aircraft and Engine	Pilots	Races
6.	**Dutch Syndicate** 7 Nieuwe Parklaan, The Hague	Pander S4 Wright 'Whirlwind'	D. L. Asjes G. J. Geysendorfer	S.
7.	**Lieut. M. Hansen** Kirkevej 2ii, Brønshøj, Copenhangen	Desoutter Mark II Gipsy III	Lieut. M. Hansen	H.

Racing No. 6 – Pander S4.

Racing No. 14 – Airspeed Courier.

Racing No.	Nominator	Aircraft and Engine	Pilots	Races
14.	**Aircraft Exchange & Mart Ltd.,** 7 Park Lane, London, W.1.	Airspeed Courier A.S.5 Cheetah V	Sq./Ldr. D. E. Stodart K. G. Stodart	H.
15.	**F/O C. G. Davies** Camden Hostel, 386 Camden Road, London, N.7	Fairey IIIF Napier 'Lion'	F/O C. G. Davies, Lieut. Comm. C. N. Hill	S. & H.

Racing No. 15 – Fairey IIIF.

Racing No.	Nominator	Aircraft and Engine	Pilots	Races
16.	**Mrs. Melrose** Glenelg, S. Australia	D.H. Moth Gipsy Major	C. J. Melrose	H.
19.	**Bernard Rubin** Berkeley Hotel, London, W.1	D.H. Comet Gipsy VIR	O. Cathcart-Jones K. F. H. Waller	S. & H.
29.	**Hospitals Trust Ltd.,** 13 Earlsford Terrace, Dublin	Bellanca Monoplane Pratt & Whitney 'Wasp' Jr.	Col. J. C. Fitzmaurice E. W. Bonar	S.
31.	**H. L. Brook** Lancaster Park Road, Harrogate	Miles 'Falcon' Gipsy Major	H. L. Brook	H.
33.	**John H. Wright** 37 Sunnyside Drive, Utica, New York	Lambert Monocoupe Warner Super-Scarab	John H. Wright	H.
34.	**A. O. Edwards** Grosvenor House, Park Lane, W.1	D. H. Comet Gipsy VIRSix	C. W. A. Scott, T. Campbell Black	S. & H.

Racing No.	*Nominator*	*Aircraft and Engine*	*Pilots*	*Races*
35.	**New Guinea Centenary Flight Syndicate** New Guinea	Fairey Fox Fairey Felix D12	R. Parer G. E. Hemsworth	S. & H.
36.	**H. C. Miller** Adelaide, S. Australia	Lockheed 'Vega' Pratt & Whitney S.C.	J. Woods D. C. Bennett	S. & H.
38	**R. C. Wallace** New Zealand	Short 'Scion' Pobjoy Niagara	S. S. Kirsten	H.

Racing No. 35 – Fairey Fox.

Racing No. 36 – Lockheed Vega.

Racing No.	Nominator	Aircraft and Engine	Pilots	Races
44.	**K.L.M.** Hofweg 9, Gravenhage, Holland	Douglas DC 2 Wright 'Cyclone'	K. D. Parmentier J. J. Moll	S. & H.
46.	**Clyde E. Pangborn** Gotham Hotel, New York, U.S.A.	Granville Monoplane Pratt & Whitney 'Hornet'	Miss Jacqueline Cochran Wesley Smith	S.
47.	**Fl/Lt. G. Shaw** Kin Shan, Nunthrope, Yorkshire	British Klemm 'Eagle' Gipsy Major	Fl/Lt. G. Shaw	H.

Racing No. 47 – British Klemm Eagle.

Racing No.	Nominator	Aircraft and Engine	Pilots	Races
50	**New York, London, Moscow, Air Lines, Incorporated** 70 Pine Street, New York City, U.S.A.	Vultee 5-1 A. Wright 'Cyclone'	Lt.-Col. G. R. Hutchinson	S.
51	**Societa Idrovolanti Alta Italia 'Savoia'**	Savoia-Marchetti S.79 Piaggio 'Stella'	Adriano Bacula Alessandro Passaleva	S. & H.
52	**R. K. Dundas Ltd.** The Airport, Portsmouth	Airspeed Courier A.S.5A Cheetah V	Mrs. Keith Miller	H.
54	**Blériot Aéronautique Société Anonyme** 3 Quai Gallieni, Suresnes, France	Blériot III Gnôme et Rhône	Capt. E. Corniglion Molinir Capt. L. Challe Vicomte de Sibour	S. & H.

Racing No.	Nominator	Aircraft and Engine	Pilots	Races
58	**Captain T. Neville Stack & S. L. Turner** Airspeed Ltd., The Airport, Portsmouth	Airspeed A.S.8 Cheetah VI	Capt. T. Neville Stack S. L. Turner	S. & H.
60	**Oliver Nicholson, President & Members of New Zealand Centenary Air Race Committee** Auckland, New Zealand	D.H. Dragon Gipsy VI	J. D. Hewett C. E. Kay	S. & H.

Racing No. 58 – Airspeed Viceroy.

Racing No. 62 – Fairey Fox.

Racing No.	Nominator	Aircraft and Engine	Pilots	Races
61	**Francis Lombardi** Vercelli, Italy	Cantieri Aeronautici Bergamaschi PL3 Fiat A59	F. Lombardi V. Suster	S. & H.
62	**J. K. C. Baines** Southdown House, Shoreham-by-Sea, Sussex	Fairey Fox Felix Curtiss	F/O. H. D. Gilman	S. & H.
63	**J. A. Mollison** Grosvenor House, Park Lane, W.1	D.H. Comet Gipsy VIR	J. A. Mollison Mrs. Mollison	S. & H.
64	**H. W. G. Penny** Sydney, Australia	Cord Vultee Wright Cyclone F2	H. W. G. Penny G. Pond	S. & H.

This information has been extracted from the official programme, published shortly before the start of the race

Appendix 2

DH 88 COMET PRODUCTION

Registration	*Constructor's No.*	*Comments*
G-ACSP	1994	First to fly, on 8th September 1934. Carried identity E-1 for 18 tests. To Portuguese Government as CS-AAJ in February 1935. To UK for restoration in 1980, but abandoned in 1982
G-ACSR	1995	Flew on or shortly before 12th October 1934. To French Government as F-ANPY in April 1935. Burnt at Istres in 1940.
G-ACSS	1996	Flew on 9th October 1934. Sole survivor. With Shuttleworth Collection. Will fly again.
F-ANPZ	1999 changed to 2260	Flew in May 1935. Built for French Government. Burnt at Istres in 1940.
G-ADEF	2261	Flew in July 1935 as E-3. Destroyed when crew baled out on 22nd September 1935.

This information is based on available evidence and is believed to be correct, but please see postscript

Appendix 3

DH COMET

PERFORMANCE ON FULL LOAD UNDER STANDARD EUROPEAN ATMOSPHERIC CONDITIONS

Dimensions	*Ft. in.*	*M.*
Span of Wing	44′0″	13.4
Height overall	8′11.3″	2.7
Length overall	29′1″	8.9
Wheel track	9′8″	2.95
Engine centres	9′8″	2.95
Mean chord	5′6″	1.68
Incidence		0° — 54′
Dihedral		3°

Areas	*sq. ft.*	*M².*
Main Planes with Ailerons	188.5	17.5
Main Planes with Ailerons and body centre section	212.5	19.8
Aileron, total	16.76	1.56
Tail Plane	11.3	1.05
Elevators, two	14.62	1.36
Fin	7.32	.68
Rudder	9.64	.9

	Speed m.p.h.	*Petrol Consumption galls/hr*	*Range with 258 Gallons (1173 litres)*
Cruising at Sea Level throttled to -3lb/sq.in. Boost	185 (298 km/hr)	18.7 (85 lits/hr)	2550 (4110 km)
Cruising at 10,000 ft (3050 m) throttled for economy	200 (322 km/hr)	17.5 (79.5 lits/hr)	2950 (4750 km)
Full Throttle at 10,000 ft (3050 m)	218 (351 km/hr)	22.0 (100 lits/hr)	2550 (4110 km)

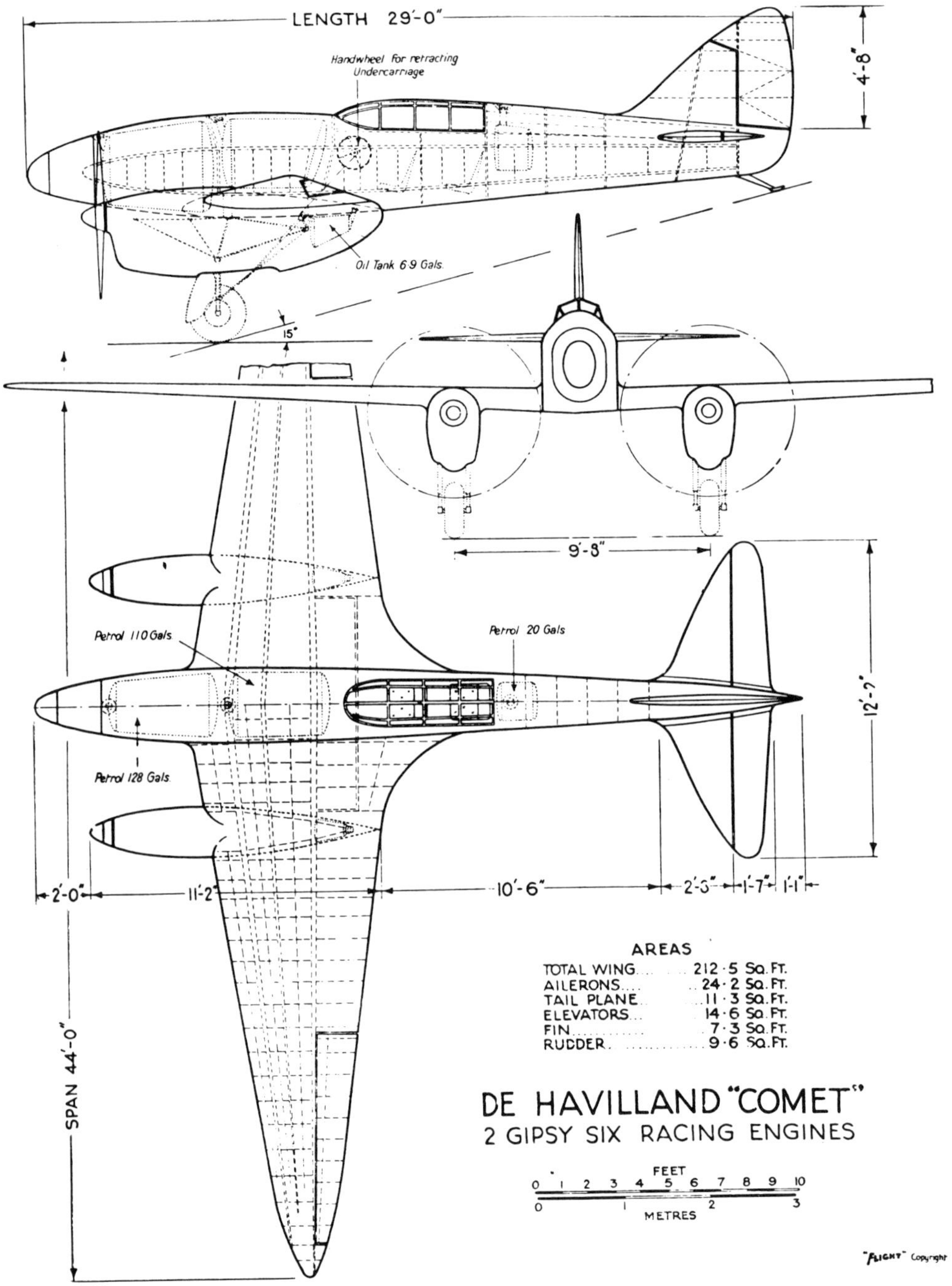
LENGTH 29'-0"
Handwheel for retracting Undercarriage
4'-8"
Oil Tank 6.9 Gals.
15°
9'-8"
Petrol 110 Gals.
Petrol 20 Gals.
Petrol 128 Gals.
12'-7"
2'-0"
11'-2"
10'-6"
2'-3"
1'-7"
1'-1"
SPAN 44'-0"
AREAS
TOTAL WING 212·5 Sq. FT.
AILERONS 24·2 Sq. FT.
TAIL PLANE 11·3 Sq. FT.
ELEVATORS 14·6 Sq. FT.
FIN 7·3 Sq. FT.
RUDDER 9·6 Sq. FT.
DE HAVILLAND "COMET"
2 GIPSY SIX RACING ENGINES
FEET
0 1 2 3 4 5 6 7 8 9 10
0 1 2 3
METRES
"FLIGHT" Copyright

Maximum Speed at Sea Level	228 mph (367 km/hr)
Take-off Run	450 yds (411 m)
Landing Run	450 yds (411 m)
Climb to 10,000 ft (3050 m)	15 minutes
Absolute Ceiling	19,000 ft (5800 m)
Absolute Ceiling on One Engine	4,000 ft (1220 m)

WEIGHTS

Tare Weight as for Melbourne Race with 258 gallons tankage (1173 litres)	3050 lbs. (1384 kgs)
Petol 258 gallons (1173 litres) at 7.7 lbs/gals	1985 lbs (902 kgs)
Oil 16 gallons (73 litres) at 9.7 lbs/gals	155 lbs (70 kgs)
Emergency drinking water	30 lbs (14 kgs)
Crew and Equipment	330 lbs (150 kgs)
TOTAL	5550/lbs (2520 kgs)

N.B. These figures were published in an internal memorandum dated 16th July 1935. They do not tally with the information on the Certificate of Airworthiness as shown on the log book entry below.

AIRCRAFT (HEAVIER-THAN-AIR).

1. Nationality British 2. Registration Marks GACSS
3. Certificate of Airworthiness No. 4540 dated 12-10-34 4. Certificate of Registration No. 5063 dated 4-9-34
5. Classification of aircraft. Category Special 6. Usual Station Hatfield
Sub-division (f) Racing or Record.
7. Owner Artur Octavius Edwards 8. Address Grosvenor House Park Lane, London, W.1. 9. Nationality British.

DESCRIPTION OF AIRCRAFT.

10. Constructor The de Havilland Aircraft Co Ltd 11. Type of aircraft D.H. 88 Comet.
12. Constructor's No. 1996 13. Maximum span (in flying position) 44'
14. Maximum length (in flying position) 30' 15. Total height (with and without trolley in the case of seaplanes and amphibians) 12.35'
16. Area of main planes 188.5 sq ft
17. Maximum total weight authorised (P)† 5500 lb. 18. Tare weight (Pv) 3002 lb.
19. Corresponding maximum total load (Pt=P—Pv) 2498 lb. 20. Minimum Crew One
21. Maximum number of persons to be carried (including crew) Two 22. Fuel Petrol.
23. Capacity of fuel tanks 257 gallons. 24. Capacity of oil tanks 13.8 gallons
25. Flying hours at international (or special) power (.6 pints per BHP hour) 15.8
26. Type and mark of engine 2x Gipsy VI R. 27. International (or special) number of revolutions of engine each 2350 28. International (or special) power of engine 2x 215/220 HP 29. Airscrews Two.
30. Apparatus for landing by land ~~and/or sea~~ Wheels. 31. Licence for wireless apparatus issued —
32. No. of Licence — 33. Type of wireless apparatus —

Given at Air Ministry, London, the 13th day of October 1934

for 34. Deputy ~~Director of~~ Civil Aviation.

† *Important Notice.*—The maximum total weight authorised corresponds to the case of the aircraft flying in dry air at an atmospheric pressure of 760mm. mercury, and at a temperature of 15° Centigrade. This weight must not in any circumstances be exceeded.

NOTE:—The Certificate of Airworthiness must be carried in the pocket attached to the inside of the cover of this book

T. 2817. Wt. 3125. 1,250 Bks. 8 34. W. & S. Ltd. (230306)

Appendix 4

FLYING THE COMET

These notes were prepared by the de Havilland Aircraft Company for use by pilots flying in the MacRobertson Air Race. Therefore much of the information provides guidance on flying for maximum range.

The notes are published as they were issued, with gaps where details were not known. The missing information was added later in the addendum printed at the end of this appendix.

D.H. "COMET" — GENERAL INSTRUCTIONS TO PILOTS

October 9th, 1934.

BEFORE FLIGHT

Ratier Pneumatic V.P. Airscrews. Immediately before flight set blades in the Fine Pitch Position by removing the plates in the front of the Spinner and pumping up the airbags to a pressure of 85 lbs. to the sq. inch (this corresponds to about   strokes with the pump). After which, the plates and springs must be carefully inspected for cleanliness and replaced without deflating the airbags by pushing them in too hard.

The blade setting is indicated on the Spinner and although this may show that the blades are in the Fine Pitch Position it is as well (in the interest of safety) to make sure that the airbags are pumped to the correct pressures.

If, prior to take-off, the correct static R.P.M. (about 2150) are not attained, the airscrews should be inspected and the airbag pressures checked to ascertain that the Fine Pitch Position has been obtained.

Engine Starting

(a) **Engine Cold.** Keep throttle closed: pull flooder wire and hold. Pull pump priming wire to bring lever into operation, and prime pump by small motion on wire until flooding is observed. Allow flooding to cease, then swing.

(b) **Engine Hot.** Close throttle and do not attempt to flood. Swing engine. If over-rich, switch off, open throttle wide and turn engine over backwards several times.

Taxying. Avoid unnecessary taxying and running up on the ground.

TAKE OFF

Both front cocks **on**
Mixture Control **off**
Flame Trap Control on "**Cold Air**".
Flaps in **second notch.**

OFF THE GROUND

Do not exceed 110 ASI. at **full** throttle, throttle back in order not to exceed maximum Permissible R.P.M.

Wind Undercarriage up a few turns to facilitate engine cooling.

Return Flap Lever to normal position. (Do not exceed 120 ASI. with flaps down).

Dive aircraft to about 150 ASI., at which speed the propeller blades will automatically set themselves in the Coarse Pitch Position, (this will become apparent by a sharp fall in the R.P.M., which may not occur simultaneously on both propellers).

Wind Undercarriage right up to "stop" and tighten knurled nut on Spindle of retracting wheel.

CLIMB

In coarse pitch at full throttle on "Cold Air" (where possible) at not less than 150 ASI. or 2000 R.P.M.

Mixture control should not be used below 7000 ft. altitude.

DURING FLIGHT

Fly when possible at full throttle at 10,000 ft.

Make sure that the undercarriage does not work down from the fully-up position.

Flame Trap. It is intended that cold air from the forward-facing air intake should be used whenever possible during the flight. This gives the greatest power output and must, in all cases, be used for take-off and initial (Fine Pitch) climb.

Hot air, taken in from the engine bay through the flame trap, should only be used if freezing is suspected.

Freezing in the carburettor is most likely to occur at low atmospheric temperatures and high humidities. It is more common with partly-closed throttle than at full throttle, but can occur under both conditions. Freezing is not necessarily worse at higher altitudes.

Freezing manifests itself in erratic running of the engine and falling R.P.M. A particular symptom is a gradual fall in the R.P.M. as the ice builds up, followed by a sudden jump in R.P.M. as a piece detaches itself and is sucked into the engine. The process is then repeated.

To eliminate freezing, close the forward air intake, by moving the lever fully rearwards into the "hot air" position. Preheated air at 30° — 40°C., will then be available at the carburettors.

In passing through a very heavy downpour, it is wise to close the forward cold air intake. The hot air intake may be used for quickly warming up a cold engine, but it is important that the lever be moved into the cold position for take-off and climb.

If the flame traps are opened to "Hot Air" some sacrifice of speed will result in exchange for an increase in range (see table below) and more mixture control will be required.

Important: Always re-set the mixture strength after any alteration to the "hot" and "cold" air lever in level flight.

Mixture Setting (permissible only at boost depressions greater than -3lbs./sq.in.)

The total fuel consumption at full throttle at 10,000 ft. should be 20 gallons per hour. This is obtained with the mixture control set to give "weakest maintained" conditions. To obtain this setting at any time, close the mixture control and fly level for a short time and obtain steady R.P.M. Open the mixture control **gradually,** watching the revolution-counter closely until a slight drop (say 25) in R.P.M. is observed. Close the mixture control **slightly,** when the original R.P.M. should be obtained. Fly level throughout this operation.

No predetermined settings for the mixture control lever can be given; this depends on atmospheric conditions, throttle opening and particularly on whether the hot or cold intake is in use. For these reasons it is necessary to reset the mixture strength at intervals (say hourly) throughout the flight, or whenever a change is made in altitude, throttle opening, atmospheric conditions or use of flame trap. Always close the mixture control and reset for mixture strength by carrying out the process as described above.

Remember, too rich a mixture will reduce your range, but too weak a mixture will result in overheating and damage to your engines.

Boost Gauge. A negative boost gauge is included to indicate when the power output is such that weak-mixture may be safely employed. It gives the pressure in the induction manifold of the port engine, and reads in lbs. per sq. in. **below** atmospheric. Thus zero boost represents atmospheric induction at sea level, and is the highest pressure obtainable.

Between zero pressure and -3lbs./sq. in. (in the red sector) the mixture control **must** be closed and in the rich position. This is the position for taxying, take-off and climb.

For manifold depressions greater than -3 lbs. (e.g. -4 or -5) the mixture may be weakened off in level flight, as described above under "Mixture Setting".

Above about 7000 ft. altitude, it will be found that the boost gauge shows greater depressions than -3lbs/sq. in. even at full throttle, and therefore full-throttle operation on weak-mixture is permissible above this altitude.

If it is desired to fly level on weak mixture at altitudes below 7000 ft. then the port engine must be throttled slightly to give a depression of greater than -3lbs. /sq. in. on the boost gauge. The starboard engine must be throttled to the same R.P.M. as the port engine.

BRIEFLY: MIXTURE CONTROL MUST NOT BE USED WHEN THE BOOST GAUGE READING LIES WITHIN THE RED SECTOR.

Range and Altitude. The Petrol consumption when flying at full throttle at 10,000 ft. at approximately 188 ASI., (255 m.p.h. true speed) and 2250 R.P.M. is approximately 20 gallons per hour or 11.25 air miles per gallon, corresponding to an ultimate range in still air of 2900 miles with full tanks (258 gallons). If greater range is required it should be obtained by flying at full throttle at higher altitudes rather than by throttling at 10,000 ft., as this latter procedure involves a considerably greater sacrifice of speed.

The table below shows the effects of Height and Flame Traps on the Range and Speed when flying with the correct mixture setting.

		Flame Traps on Cold Air			*Flame Traps on Hot Air*		
Altitude Feet	*Condition*	*Range*	*ASI.*	*True Speed*	*Range*	*ASI.*	*True Speed*
14,000	Full Throttle, weak						
12,000	Full Throttle, weak						
10,000	Full Throttle, weak						
8,000	Full Throttle, weak						
7,000	Full Throttle, weak						
5,000	Full Throttle, rich						
Sea Level	Full Throttle, rich						
5,000	Throttled to -3lbs./sq.in. Boost, weak						
Sea Level	Throttled to -3lbs./sq.in. Boost, weak						

DURING FLIGHT (cont.)

The following important points emerge:

(1) Fly at 10,000 ft. at full throttle under normal circumstances.
(2) Fly higher at full throttle for increased range.
(3) Flight at low altitudes throttled to -3lbs. boost is uneconomical in speed and range.
(4) Flight at low altitudes at full throttle is very uneconomical in range.
(5) The use of "hot" air decreases the speed, though increases the range.

Notes

1. **Mixture Control.** Failure to use any mixture control at 10,000 feet would result in a reduction of the range from

2. **Throttling.** The relative ineffectiveness of throttling for range as compared with flying high may be gauged from the fact that if at 10,000 feet the throttle is closed sufficiently to give the same range as would be obtained at full throttle at 14,000 feet the true speed would fall to instead of m.p.h.
3. **20 m.p.h. Following Wind.** For every hour during which this is experienced, the range will be increased by 20 miles; and conversely for a head wind.

Petrol Consumption and Trim. The aircraft will tend to become tail heavy towards the end of the flight, and it must be arranged that when the remaining petrol falls to about 70 gallons, this should all be in the front tank. In no circumstances must the rear tank be full when the front tank contains less than 70 gallons.

The following order of feed is recommended:-

1. Exhaust centre tank completely. (The time occupied and distance covered in doing this afford a useful check on the range).
2. Use 50 gallons from front tank.
3. Exhaust rear tank.*
4. Exhaust front tank.

LANDING

The C.G. is in its fullest aft location when landing with almost empty tanks.
Undercarriage down. (Wind hand wheel until red light on dashboard goes out and 'Stop' is reached).

Flaps full down in 3 stages at a speed not exceeding 120 m.p.h. The last stage should occur at gliding speed (95 ASI.) Flaps down makes aircraft trim tail heavy.

Glide in at 85-95 ASI.

* The rear tank will not feed when the front tanks are turned on, therefore turn off front cocks when feeding from rear tank.

ON THE GROUND

When the aircraft lands the propeller blades will naturally be in the Coarse Setting, and prolonged full throttle taxying on boggy aerodromes should be avoided by changing the pitch to Fine, which should again be checked immediately before take-off.

GENERAL NOTES

Engine Revolutions. The figures below are given as a guide as to what may be expected under different full-throttle conditions:

Condition	*Static*	*Fine Pitch Climb*	*Coarse Pitch Climb*	*Level at 10,000 ft.*
Approx. A.S.I. R.P.M.	- 2150	110 2500	150 2000	188 2250 Cold Air 2150 Hot Air

The terms "hot air" and "cold air" refer to whether the carburettor intake is through the flame trap from the vicinity of the cylinders ("hot") or through the external forward-facing intake ("cold").

Fuel. Fuel and oil specifications must be rigidly adhered to. The fuel is Anglo American Oil Company's D.T.D. 224.

Oil. Wakefield's Castrol XXL.

Approximate consumption 3¼ pints per hour. Tank capacity 6.9 gallons. Replenish at each stage of flight.

Oil pressure 40-45 lbs./sq. in. Clean filter at each stage of flight by rotating handle one or more turns using slotted spanner provided. (It is not necessary to remove the cowling for this purpose.)

Sparking Plugs. If, for any cause, the engine cowling is detached during the Race, or if, for some reason, the machine is held up for a time, the opportunity should be taken for changing to a new set of K.L.G. V.12/1 Sparking Plugs. Otherwise, unless definite plug trouble develops, the original set may be used throughout the Race.

The emergency 3-gallon drinking water tank and the accumulator are inside the fuselage nose cowl.

ENGINES RUNNING WELL — LEAVE WELL ALONE

Flight on One Engine. Rudder Trim gear is provided on control box below front pilot's seat.
Tyre Pressures. 32lbs/sq. in.
Petrol System. A diagram with explanatory notes and instructions is attached. The front tank should always be filled last. The cock for the rear tank must remain "off" until level flight is attained, otherwise the petrol in the front tanks will drain back and overflow from the vent of the rear tank.

If for any stage of the race, it is desired to leave off petrol, this should be left out of the centre tank, the other two being filled full.

GENERAL NOTES (cont.)
General Loading Instructions for Practice Flights. In order to keep the centre of gravity of the aircraft within the prescribed limits, it is desirable that the following loading instructions should be observed when filling up the tanks for flight.

Drinking water tank — full
Oil tanks — half full to full
Centre fuel tank — 5 gallons

			Petrol in Front Tank *Not More Than*	Petrol in Front Tank *Not Less Than*
SOLO Pilot in Front Seat	Rear Tank Empty	Flying Weight C.G.	25 galls. 3500 lbs. Forward	- - -
	Rear Tank Full	Flying Weight C.G.	75 galls. 4100 lbs. Forward	25 galls. 3700 lbs. Aft
DUAL	Rear Tank Empty	Flying Weight C.G.	75 galls. 4100 lbs. Forward	25 galls. 3700 lbs. Aft
	Rear Tank Full	Flying Weight C.G.	Full 4650 lbs. Forward	70 galls. 4200lbs Aft

The quantity of petrol in the centre tank is unimportant but should be kept low in order to keep the total weight down.

During the Race the aircraft will be taken off on an all-up weight of approximately 5500 lbs. and a fairly forward C.G.; and landed on a weight of approximately 3500 lbs. to 4000 lbs. with aft C.G.

BRIEF INSTRUCTIONS FOR PILOT

BEFORE TAKE-OFF.
Fuel & Oil Tanks Full, Airscrews in Fine Pitch.
TAKE-OFF.
Petrol on, Flaps half down, "Cold Air"
AFTER TAKE-OFF.
Flaps Up, Undercarriage Up, Propellers in Coarse Pitch.

FLIGHT.
Use Mixture Control, Check Undercarriage Up, "Cold Air", Boost, Petrol Feed and Trim.
LANDING.
Undercarriage Down, Flaps fall down gradually.

D.H. 'COMET' — ADDENDUM TO PILOTS' INSTRUCTIONS

The completed table on page 3 is as follows:

		Flame Traps on "Cold Air"			*Flame Traps on "Hot Air"*		
Altitude ft.	*Condition*	*Range*	*ASI*	*True Speed*	*Range*	*ASI*	*True Speed*
14,000	Full throttle, weak	3150	171	216	3290	161	204
12,000	Full throttle, weak	3020	181	221	3165	170	208
10,000	Full throttle, weak	2880	190	225	3030	179	212
8,000	Full throttle, weak	2730	199	228	2890	188	216
7,000	Full throttle, weak	2660	204	230	2825	193	218
5,000	Full throttle, rich	1720	213	233	1720	202	221
Sea Level	Full throttle, rich	1760	239	239	1760	230	230
5,000	Throttled to -3 lbs. sq. in. boost, weak	2690	203	222	Same	as	on
Sea Level	Throttled to -3 lbs. sq. in. boost, weak	2770	201	201	"Cold	Air"	

The above figures are approximate only and should be used merely as a guide to the relative effects of the different variables.

The speeds are the mean speed throughout the flight — the actual speeds would be lower than the above at the beginning of the flight when the aircraft is heavy and higher at the end. The value of flying above 10,000 ft. for increased range becomes much greater as the weight gets less. It is recommended that the aircraft be not flown above 10,000 ft. for the first 3 hours of the flight.

The economy of flying high will be lost if the aircraft is flown at such a great height that the full travel of the Mixture control lever fails to give any RPM. drop.

In the event of the boost gauge failing to function, the following approximate RPM. may be used as an alternative guide to the degree of throttling to be used when running weak below 7000 ft.

HEIGHT **R.P.M.**

As is apparent from the table flight at full throttle below 7000 ft. is very uneconomical and should only be indulged in in the case of emergency when range is unimportant. In such a case it is both faster and more economical to fly near ground level.

Try Magnetos whilst approaching the aerodrome — this may save time on the ground.

The blank space in **Note 1 on Page 4** should read "from 2900 miles to 2200 miles."

The blank space in **Note 2 on Page 4** should read "to 205 mph instead of 216 mph."

Appendix 5

FLIGHT TEST REPORTS

Summary of flight tests from reports held in DH/BAe Hatfield aerodynamic archives

No detailed reports of the first three flights have been found, except for the summary on the page for Test 1 (13/9/34) which was the 4th test flight. This catalogue begins with Test 1.

Test 1 — Thursday, September 13th, 1934. 4th Test Flight, 3900 lbs.
Modified forward-facing intakes in nose of cowling
Pitch adjusted since first flights
Tailplane setting 2° negative
U/C position doubtful

Test 2 — Sunday, September 16th, 1934, 5th Test Flight, 4063lbs
Tests on side intakes — not totally successful, intake flap stuck

Test 3 — Sunday, September 16th, 1934
As Test 2, but with flap wired open to side intake

Test 4 — Monday, September 17th, 1934, evening
As Test 3

Test 5 — Tuesday, September 18th, 1934, morning
As Test 4 but with flap wired open to flame trap & inside intake, Static leak discovered in A.S.I. — Instrument changed for one with a slower static leak

Test 6 — Tuesday, September 18th, 1934, 4063 lbs.
Speed course at 1000 ft. G.L. if possible
Left engine rough at ¾ throttle — gradually improved
Speed course at 10,000 ft. impossible due to cloud.
Speed course at low altitude
Luton Railway Station — Royston Railway Station & return — mean speed 222 mph

Test 7 — Thursday, September 20th, 1934
Engines taken down and modifications made to piston rings since last flight
Flap travel extended to 47° from 35°
Longer tail skid added

Test 8 — Tuesday, September 26th, 1934, 4095 lbs.
First flight with Ratier props — not correct spinners yet.
Props adjusted to give 2200 R.P.M. static
Left prop change at 110 ASI, Right prop change at 160 ASI

Test 9 — Tuesday, September 26th, 1934, 4595 lbs.
Machine as Test 8, except that props set to give 2250 R.P.M. at 10,000 ft.
Purpose: Screen take off, consumption & timed course at 10,000 ft., fuel dumping.
Speed run not possible due to cloud
Fuel dumping — 30 gals. in 21 minutes from front tank
Flaps could not be operated beyond 35°
Screen take off — 75 ft. at Screen, 12 mph. wind

Test 10 — Thursday, September 28th, 1934, 4595 lbs., (55 minutes)
Machine as Test 9, except correct spinners fitted
New tail skid etc.
Oil consumption Test

Stall	4800ft	Engine off	Flaps Up	88 ASI
Stall	4800ft	Engine off	Flaps 1st Notch	86 ASI
Stall	4800ft	Engine off	Flaps 2nd Notch	84 ASI
Stall	4800ft	Engine off	Flaps 3rd Notch	79.75 ASI

Test 11 — Thursday, September 28th, 1934, (10 minutes)
As Test 10, but with plates removed from front of props. to prevent pitch change
Stall; 1500 ft. Engine on, Chassis down, 82 ASI

Test 12 — Friday, September 29th, 1934, 4795 lbs., (45 minutes)
Purpose — speed course at 10,000 ft. to Farnborough
Left hand engine "packed up on speed course at 10,000 ft"

Test 13 — Sunday, October 1st, 1934, 10 a.m., (90 minutes), 4095 lbs., approx.
New L.H. engine fitted — Larger air scoops; louvres aft on both sides,
Purpose: top speed determination
Speed course 10,000 ft. Hatfield to Brookwood Railway Station and return — 224.1mph.
Speed course 5000 ft. Luton Railway Station to Royston Railway Station and return — 230.5mph

Test 14 — Sunday, October 1st 1934, 6 p.m., 4095 lbs.
Purpose: Practice take-offs with and without flaps.

Take-off	No. 1	No flaps	10.4 secs.	Hedge 26 secs.
Take-off	No. 2	1st Notch	10.8 secs.	Hedge 26 secs. Worse
Take-off	No. 3	2nd Notch	10.2 Secs	Hedge 26 secs. Better than either

Test 15 — Monday, October 1st, 1934, 4695 lbs.
4 Practice take-offs, Flap 2nd notch, Screen speed 90 mph.

Test 16 — Friday, October 5th, 1934, 5109 lbs., 9.30 a.m.
Official Take-off and landing tests
Observed by: Mr. Young (Resident Technical Officer)
Flt. Lt. Boothman
Flt. Lt. Pegg
Pilot: Capt H. Broad
Wind 7 mph., for take-off, 12½ mph., for landing
Take off: 295.4 yards to unstick, 116 ft. high at 749 yard screen
Landing: 299 yards ground run

Test 17 — Friday, October 5th, 1934, 4650 lbs.
Climb & Consumption with flow meter
Very, very surgy weather — results very unreliable

Test 18 — October, 1934, 5100 lbs.
General Handling Trials
All seemed satisfactory

Test 19 — Thursday, October 11th, 1934, 3pm, 4250 lbs.
Same machine as previous tests but finished & **painted green** *
Consumption tests

Test 20 — Friday, October 12th, 1934, 10 a.m., 4250 lbs.
Same machine as above
Consumption tests and climb performance
"Both props. went off at 160 ASI"

Test 21 — Friday, October 12th, 1934
Machine as above
Consumption tests — larger jets
Bumpy — ASI unsteady

Test 22 — Mildenhall Red Machine, 4200 lbs. approx
Scott & Black

Test 23 — Mildenhall
Cathcart-Jones

Test 24 — 12/12/34 to 18/12/34 — G-ACSR
Engine Tests, Test log EE 279

Test 25 — Memo dated January 7th, 1935 — G-ACSR
Engine cooling Tests

Test 26 — Memo dated January 11th, 1935 — G-ACSR
Repeat of Test 25 due to incorrect pressure readings

Test 27 — Memo dated January 14th, 1935 — G-ACSR
Repeat of Tests 25 & 26 with modified cowling

Test 28 — Memo dated January 19th, 1935, Test: 14th January, 1935
D.H. 82 Tiger Moth comparative cooling tests
(1st page only of memo, subsequent page(s) missing.)

Test 29 — Memo — first page missing! no date — G-ACSR
Depressions on leading edge of plane

Test 30 — Memo dated January 21st, 1935 — G-ACSR
Effective pressures for engine cooling and depressions on leading edge of plane, (also test 30a)

Test 31 — No date DH 89 Dragon Six/Rapide
Comparative cooling tests

**This conflicts with the logbook evidence printed on page 26, but see postscript.*

Test 32 — January 17th, 1935 (memo dated 22/1/35) G-ACSR
as tests 29 & 30 etc.

Test 33 — Memo dated January 21st, 1935 — G-ACSR
Effective pressures for engine cooling

Test 34 — Memo dated January 21st, 1935 — G-ACSR
Repeat of Test 33 with cowling in original condition

Test 35 — February 23rd, 1935, CS-AAJ, 4290 lbs.
Fuel consumption tests

Test 36 — March 1935 For French Air Ministry, 5000 lbs. approx
15 (or 16) test flights in all, all called test 36
Engine failure on 4th flight
Max speed 218.5 mph, 10th flight

Test 37 — no date given, 4400 lbs. approx
Consumption tests

Test 38 — March 1935 — Pilot: Capt. Broad
New aft. C. of G. position

Test 39 — April 8th, 1935, 5000 lbs.
Comet for French Air Ministry
Single engine performance

Test 40 — May 2nd, 1935 (date of report)
DH Comet for French Air Ministry
Results of acceptance Tests for First Machine — F-ANPY
Tested at Villacoublay
(Flown by H. Buckingham)

Test 41 — May 31st, 1935, Machine 2nd French Comet, 1999
Second Machine for French Air Ministry
To check jet settings adopted for previous "Green Comet"
Engine log EE 322

Test 42 — May 1935, 5000 lbs. approx
2nd Machine for French — not finally painted
Speed course: Hatfield-Reading & return — 223.4 mph

Test 43 — May 31st, 1935
2nd Machine for French A.M.
Speed course: Luton-Royston & return
Cruise at 180 ASI — Speed 183.5 mph.
Cruise at 200 ASI — Speed 204 mph.
Cruise at Full Throttle — Speed 228 mph.
All at 1200 feet.
There follows a note about the acceptance tests at Villacoublay, July 1935, for the 2nd French Comet taken from a letter from a Mons. Beauregard, July 17th, 1935.

RPM	SPEED
2335	226.5 mph.
2020	193.2 mph.
1495	141.0 mph.

(Actual observed timed max. speed = 233 mph.)

Test 44 — 3rd June, 1935
Timed consumption tests at 10,000 feet
Speed 197 mph — rage 2340 miles

Test 45 — June 6th, 1935, 5000 lbs. 2nd French Comet, 1999
Apparently a continuation of Test 41 — 4 more flights
Engine test report EE 322, Schedule No. 115

Test 46 — June 27th, 1935
Check of carburettor settings for French A. M.
A/C DH 88m K 5084
Engine test report EE 328

Test 47 — This has been allocated to the results of acceptance tests at Villacoublay, July 1935 for the 2nd French Comet.
Take off (unstick) — 220m
Height at 750m — 32m
Landing Run — 400m
Speeds — see test 43 etc.

Test 48 — July 30th, 1935, 5th machine (Mr Campbell Black)
Check mixture at 10,000 ft.
Check consumption at -3.0 lbs./sq. in. boost
6 flights — 1st five — Capt. Broad
6th — Mr Waight
Observer — Mr. J. Walker

Test 49 — Memo dated August 9th, 1935, 5635 lbs.
Mr. Campbell Black's take-off, 509 yards to unstick, Wind 5 mph.

Test 50 — August 10th, 1935, 5579 lbs. K5084 (late G-ACSS)
Full load tests on Air Ministry machine
Corrected t/o to unstick — 402 yards, Wind 5mph.
Stall Flaps up 92 mph.
Stall Flaps down 84 mph.
Memo dated 12/8/35 plus report by Mr. Waight 10/8/35
Pilot: Mr. Waight — Observer: Mr. Herrod-Hempsall
Problems experienced keeping aircraft straight.

Test 51 — August 8th, 1935, 5615 lbs. Machine 1994, CS-AAJ
Reconditioned after crash, otherwise in similar condition as for Race
Purpose: Climb to service ceiling, level speeds at altitude, propeller characteristics at full throttle & cruise in coarse pitch
Pilot — Mr. Waight; Observer — Mr Herrod Hempsall
"In taking-off there was great difficulty in getting the tail up, with a tendency to swing to the right"

Test 52a — September 8th, 1935, 5000 lbs. CS-AAJ
Pilot — Mr. Waight; Observers — Mr Clarkson & Mr. Herrod Hempsall
Fuel consumption at 1000 ft. to check throttled specific consumption

Test 52b — September 9th, 1935
Continuation of 52a

Test 53 — September 11th, 1935, 5000 lbs. G-ADEF, Machine 2261
Pilot — Mr. Waight; Observer — Mr. Walker
Climb to 10,000ft., coarse pitch, full throttle, cold air.
Full throttle speed at 10,000ft., hot air — 212 mph } T.A.S.
Full throttle speed at 10,000ft., cold air — 224 mph } T.A.S.
Fuel consumption — 18.09 gph. range 3020 miles
Oil consumption — 3.47 pts. p.h.

Test 54 — September 19th, 1935, M/C 2261 G-ADEF
Repeat of Test 53 plus check on petrol & oil consumption

Test 55 — January 7th or 8th, 1936, K5084, 4876 lbs.
Trim and Fore and Aft Stability Tests on an intermediate C. G.
Pilot:- Mr Waight.
Compared with DH86A longitudinal stability
Purpose of test was apparently to obtain design data for another de Havilland type — believed DH97
This concludes all the Test Reports discovered.
Also in the archive is a report on the flight test of Leopard Moth, G-ACHD (prototype) fitted with Gipsy VI R on 18th June 1934. Test number 15 (a — k (excluding i)) it is not stated whether this is test 15 in the engine certification series or in the Leopard Moth series

SUMMARY OF COMET MAX. SPEEDS & PETROL CONSUMPTION
This summary is an average of all the tests performed
Petrol consumption at full throttle at 10,000 ft. — 22 gph.
Petrol consumption throttled to mark (Hot Air) at 10,000 ft. — 17.5 gph.
Cruising speed throttled to mark (Hot Air) at 10,000 ft. — 198.5 mph. (Hot Air)
French Comets — 201.5 mph. (Cold Air)

	Race Comets	French Comets
Maximum speed at 10,000ft.	222.5 mph.	218 mph.
Maximum speed at sea level	235 mph.	228 mph.

N.B. This summary has been taken from the Hatfield records. These are handwritten by Richard Clarkson, mainly in pencil. Some dates mentioned do not tally with the days of the week and some entries differ by a day or so from those in Hubert Broad's logbook. However, readers should remember that the entire test programme was carried out in a very short time. From the first Comet's first flight to the start of the race only six weeks elapsed, so everyone concerned was under great pressure.

Appendix 6

ORGANISATIONS THAT HAVE PROVIDED TECHNICAL OR MAJOR FINANCIAL SUPPORT FOR THE COMET RESTORATION PROJECT

Ansett Airlines
The Army Aviation Centre
The Army Parachute Centre
Bestobell Aviation Products
Bowden Controls
Bowker Air Services
British Aerospace — Aircraft Group
British Aerospace — Dynamics Group
British Airways
British Insulated Callender's Cables
Bruntons (Musselburgh)
Henry Browne and Son
Ciba-Geigy
The Civil Aviation Authority (Airworthiness Division)
Cobaircraft
CSE Aviation
Dowty Group
Dufay-Titanine
Dunlop Aviation Division
Charles Elphick and Son
Field's Aviation Accessories
Flight International
GQ Parachutes
Graviner
Hants and Sussex Aviation
ICI Plastics
Irvin (Great Britain)
Letraset (UK)
Litco Flight Aids
Lucas Aerospace
MacMillan Bloedel
Manganese Bronze
Marconi Avionics
Marshall of Cambridge
Mingay's (Chigwell)
Mollart Engineering
Neraglow
Qantas Airways
Plessey Aerospace
Renold and Coventry Chain
RHP Aerospace
Rolls-Royce
Rose Bearings
Rosemount Engineering
The Royal Aircraft Establishment
Scotney
Smiths Industries
Sterling Metals
Specialised Mouldings
Sperry Flight Systems
Staravia
Superflexit-Icore
Symonds Distributors
Teleflex
The Transport Trust
Triplex Safety Glass
Henry Wiggins and Son

DH88: Bibliography

British Racing and Record Breaking Aircraft; Peter Lewis

DH: A History of de Havilland; C. Martin Sharp

Florence Desmond – by herself; Florence Desmond

Scott's Book; C. W. A. Scott

The Great Air Race; Arthur Swinson

(Also *Men with Wings,* but I cannot trace the name of the author as I had only three photocopied pages from an unknown source)

Items of information have come from many sources, written and verbal, but these are the books from which details have been taken and/or facts cross-checked.

Postscript

As the book has developed, some details that I have sought and obtained from a wide variety of sources have led to increases in the amount of conflicting information. Weights and performance figures differ from one source to another, but some of this may be understandable because of other variables; for example, the maximum authorised weight on the original Certificate of Airworthiness differs from the figure at take-off for the race because the Resident Technical Officer from the Air Ministry very sportingly allowed the limit to be increased only two days before the start; also, various tests indicate an assortment of figures for the time taken to climb to 10,000 feet, but these can be of value for comparison purposes only if the weights are known, for clearly a lightly loaded Comet and one with all tanks full would be virtually two different aeroplanes. Another point to be considered is that the performance on early tests with the original Hamilton propellers might have been noticeably different from that with Ratiers, while a few flights with the latter were carried out with the blades fixed in fine pitch and (perhaps surprisingly) the normal instruction to Comet pilots was to climb in coarse.

The most difficult problem to solve is in the search for proof of which was the first DH 88 to fly. Hubert Broad's logbook and most other available evidence lead to G-ACSP (entered initially as No. 1994), but the reference on page 164 to 'same machine as previous tests but now finished and painted green' must relate to G-ACSR; yet this refers to a date that is one day before Hubert Broad logged the machine's first flight! Also, this is entered as being of 3 hours and 25 minutes duration, which is most unlikely; the first flights of 'CSP and 'CSS were of 25 and 20 minutes respectively. At one stage I looked at the likelihood that the constructor's numbers had been confused, but I have found one Air Ministry and three de Havilland documents that state G-ACSP as 1994 and two that quote G-ACSR as 1995; also, a memorandum on a DH file mentions G-ACSR/1995 and G-ACSS/1996 as the second and third machines respectively. Another angle that I investigated was the possibility that another pilot may have carried out the first flight or flights of 'CSR, but I have received categorical confirmation that Hubert Broad conducted all the initial tests.

I can think of only two possible solutions: that Hubert Broad flew G-ACSR on several flights before 12th October 1934, but in the rush of the time either he forgot to enter them in his logbook or he grouped these trips together to make the total of 3 hours and 25 minutes and recorded them all for that date. As the average duration of the first eleven recorded test flights was only 33 minutes, this becomes almost as much a probability as a possibility.

Perhaps we will never know the truth. Certainly I have exhausted all the information sources that I can find, but someone, somewhere, may have evidence to prove that G-ACSP took to the air before G-ACSR — or *vice versa.* If you are that person, please come forward. This may be only a minor point in the total tale of the Comet, but I have aimed to find and record as many relevant facts as possible and this is one that has escaped.

'Grosvenor House', almost complete again, on the first day of the Farnborough display in September, 1984. (Austin Brown).

Index